Email – SMS – MMS

Linguistic Insights

Studies in Language and Communication

Edited by Maurizio Gotti,
University of Bergamo

Volume 58

PETER LANG
Bern · Berlin · Bruxelles · Frankfurt am Main · New York · Oxford · Wien

Carmen Frehner

Email – SMS – MMS

The Linguistic Creativity of Asynchronous Discourse in the New Media Age

PETER LANG
Bern · Berlin · Bruxelles · Frankfurt am Main · New York · Oxford · Wien

Bibliographic information published by Die Deutsche Bibliothek
Die Deutsche Bibliothek lists this publication in the Deutsche Nationalbibliografie; detailed bibliographic data is available on the Internet at ‹http://dnb.ddb.de›.

British Library Cataloguing-in-Publication Data: A catalogue record for this book is available from *The British Library,* Great Britain

Library of Congress Cataloging-in-Publication Data

Frehner, Carmen, 1978-
Email, SMS, MMS : the linguistic creativity of asynchronous discourse in the new media age / Carmen Frehner.
p. cm. – (Linguistic insights, 1424-8689 ; v. 58)
ISBN 978-3-03911-451-1 (alk. paper)
1. Discourse analysis. 2. Telematics. 3. Creativity (Linguistics) 4. Register (Linguistics) 5. Communication and technology. I. Title.
P302.865.F74 2008
302.23'1–dc22

2008003372

Accepted in the winter semester 2006/07 on the recommendation of Prof. Dr. Udo Fries and Prof. Dr. Christa Dürscheid.

ISSN 1424-8689
ISBN 978-3-03911-451-1

Hochfeldstrasse 32, Postfach 746, CH-3000 Bern 9, Switzerland
info@peterlang.com, www.peterlang.com, www.peterlang.net

Printed in Germany

To my parents

Table of Contents

Preface

– sum –
– nite – gud –
– tho – cn – tb –
– oda – rly – 2moro – soz –
– tym – nel – hapi –
– sed – dat –
– nofin –

These few words represent only a very short list of examples of the great variety of spelling found in the emails, text and multimedia messages analysed for this study. Some of the words may be easily comprehensible while others, such as the word *cn*, which has two meanings, namely 'can' or 'seen', seem more difficult. Flipping through these words, one might get the impression of a completely irregular orthography – randomly spelt words having nothing to do with what was learnt at school. Custodians of the English language may get wrinkles and grey hair upon looking at such a word list and wonder where this may lead us. Yet, reality is not as grim. In fact, all of the above words follow certain rules – not one of the lexical items is randomly spelt. Even though these rules are not described in any grammar books, they appear to be of general value. People are aware of them and follow them when composing emails, text messages or multimedia messages.

So far, these rules have not been described in any study as there is generally only very little research material available on the linguistics of private emails, text messages and multimedia messages. What is more, a great majority of the studies that are available have to be handled with great care, for they were conducted either in non-Anglophone areas or used messages that were written in an economic context, i.e. in official and formal environments. These criteria must not be underestimated; they eventually make it difficult to compare earlier research with the data of this book, which covers private emails, text messages and multimedia messages. Nevertheless, the opportunity to pursue a research project in a field that has thus far received little academic interest has been highly promising and brought many crucial results to light.

To be able to carry out such a revealing study, I was dependent on a helpful and accommodating environment and I would like to thank all of those who have supported me in this project and fundamentally contributed to the development of this book. I am most grateful to Prof. Dr. Udo Fries for supporting and encouraging my work in all its aspects as well as to Prof. Dr. Christa Dürscheid, whose well-founded and deep knowledge of the subject was a great enrichment. The people who have helped with the data-collection are too numerous to list, but a special thank you must go to Richard Abraham. Many thanks also to Susanne Böttner for doing the proofreading and a very big thank you to my parents for their unfailing moral support and encouragement throughout the project for which I am highly appreciative.

List of Abbreviations

ARPANET	Advanced Research Project Agency Network
ASCII code	American Standard Code of Information Interchange
BBoard	Bulletin Board
CMC	Computer-Mediated Communication
CMD	Computer-Mediated Discourse
Email	Electronic Mail
GSM	Global Standard for Mobile Communication
HTML	Hypertext Markup Language
IM	Instant Messaging
MMS	Multimedia Message Service (also short for 'multimedia message(s)' and 'picture message(s)' respectively)
MSN	The Microsoft Network
SMIL	Synchronised Multimedia Integration Language
SMS	Short Message Service (also short for 'text message(s)')
T9	Text on nine keys
URL	Uniform Resource Locator
WAP	Wireless Application Protocol

1 Introduction

1.1 The Topic: A General Survey

> oi dnt feel bad bec of me.Its just dat u knw God has taken so many ppl dat were so close 2 me.Dat I just h8 life.Bt no mata hw mch pain or h8 I am in I wil neva shw it on my face I kep it 2 myself.N like I said I cry from my heart and not from da eyes. But lok.u smile dnt wory abt me.I wil be happy.K.Cheerup.

Whereas nowadays private letters, messages and notes are often type-written, there was a time when personal letters were never typed but always handwritten because "type-written communication was reserved for and associated with business correspondence" (Meyers 1994: 4). It used to be much more personal to write them by hand as "a note received in your own handwriting seems to convey your feelings much more personally than one that is written on a word processor or type-writer" (Meyers 1994: 4). With the advent of email[1] and text messages (SMS), however, private letters and messages are no longer exclusively handwritten but also typed. Thereby, and not only from the nostalgic point of view of traditionalists, it is often regretted that these letters and messages have lost a certain personality and individuality, which, of course, cannot be denied: whereas the appearance and legibility of a handwritten letter as well as its writing paper make a non-electronic letter personal, these cues are not to the same extent present in emails,

1 Even though Jansen's *NetLingo Internet Dictionary* postulates the hyphenised spelling of the word email, i.e. "e-mail", arguing that email is short for "electronic mail", it will be spelt without a hyphen in this volume for the simple reason that the language of CMC tends to be reduced, so hyphens do not fit in the concept of shortenings. Also, it can be recognised that people tend to spell it unhyphenated when examining the corpora of this study: of the 208 instances of the lexeme email throughout the corpora, 189, i.e. 91%, are spelt without a hyphen whereas only 19 are hyphenated.

text messages or multimedia messages (MMS). It is therefore necessary to incorporate linguistic features in these messages to make up for this loss. Furthermore, even though email, SMS and MMS are asynchronous media that make direct interaction impossible, they have the potential to approximate synchronicity and enable near-real-time written conversation to such an extent as it has never been known before. Therefore, and since less information can be conveyed in the same amount of time by writing than by speech, the language has to be adapted to the new media to make written conversation more efficient, less time-consuming and nevertheless comprehensible so as to keep misunderstandings at a minimum. And the language has indeed changed: with the increasing use of email, SMS and MMS, new registers have developed, namely hybrid registers that mix the features of the spoken language with those of the written language which often results in a cryptic, encoded and hieroglyphic language of the following type (1):

(1) hey babe soz bout earlier phne cut out n cnt find d charger.Txtnu frm [name] nw.speak2u2mz hope u have a nice sleep widout ne interuptions2nite.coz i cnt tak it wenur crnky cz u dnt gt enuf slp.Love u ur bf xxx<3

The aim of this book is to characterise these new registers and to show that they are, while being composed in a written medium, gradually shifting towards (or even beyond) the spoken language[2]. Being in midst of this process, however, the rules and conventions of these registers are still evolving; there are neither rights nor wrongs when composing messages in these media. Hence, the data will not provide a definitive characterisation of the text types[3] email, SMS and MMS, which, in any case, would be impossible: as long as "computer networks and computers

2 In this study, the term "spoken language" only refers to unplanned discourse such as face-to-face or telephone conversation. Planned discourse (e.g. written or educated speeches or interviews) is explicitly excluded.

3 When using the term text type in this context, I do not refer to emails, SMS or MMS in general but specifically to the text type of private emails, SMS or MMS which are written among students between 17 and 27 years of age. Within this category, I do not further differentiate between any subcategories such as greeting messages or goodnight messages, but subsume them under the same category or text type, respectively: private messages among students.

themselves become even more powerful" (Yates 1996a: 82), the language of emails and SMS will continue to change. Rather, I will point out certain tendencies that can be established when examining electronically composed messages. The book will first look at previous research in this field and describe the underlying methodology of this study. After a broad overview of the register of emails and text messages, it will offer a close analysis of these text types, their historical background and characteristic linguistic features as well as a comparison and more accurate placement of email and SMS along and beyond the written/spoken continuum. Koch and Oesterreicher's model (1985, 1994) will offer a good means to better understand the linguistics of email and text messages. It will further be shown in what way the model has to be extended to actually fit the new communicative conditions. Email and text messages will then be compared to letter mail and telephone conversation, while a close comparison of SMS and telegrams will show to what extent it is legitimate to speak of a renaissance of telegrams when it comes to text messaging. Subsequently, attention is turned to multimedia messaging and photo messaging respectively. Since it was impossible to collect an adequate corpus in the United Kingdom, examples from the German and Swiss-German corpora will be consulted. In contrast to emails and SMS, the focal point will not be on the hybrid register. Instead, questions concerning the proportion of image to text, picture categories as well as MMS dialogues will be examined in this section.

Since there have been numerous complaints from various sides claiming that texting is leading to the deterioration of language, I shall also comment on the linguistic changes and their implications, which then leads to another complaint, namely one within the German language, concerning the increasing use of Anglicisms. To see whether and to what degree electronically composed messages indeed foster the use of Anglicisms, the German and Swiss-German email and text message corpora will be analysed in terms of Anglicisms. Finally, I shall draw some preliminary conclusions and speculate about possible future tendencies.

1.2 Previous Research on Emails, Text Messages and Multimedia Messages

Höflich and Rössler (2001: 445) remark that relatively little is known about the private use of email even though it is one of the most popular Internet services; there has indeed not been much research on private email correspondence until today. This is not only due to the fact that email is a recent communication medium, but also to problems in gathering data. Most previous studies on email conversation have focused on the use of email as an official means of communication in business because "computer-mediated communication [CMC] was first established to transact business" (Huff and Rosenberg 1989 quoted from McCormick and McCormick 1992: 388, my comment); for this reason there has been little academic interest in the private use of email.

As early as 1984, Naomi Baron published an article on "computer mediated communication as a force in language change", but "it was not until 1991, with the publication of Kathleen Ferrara, Hans Brunner, and Greg Whittemore's 'Interactive written discourse as an emergent register', that linguists and language scholars began to take serious notice of CMD [computer-mediated discourse]" (Herring 2001: 613, my comment). The year 1991 can be considered the key year of CMC-research; it is this the year in which linguistic interest in computer-mediated communication first emerged. In the same year, Kiesler and Sproull published a book in which they examined to what extent new computer-based communication technology is changing the way people work. While many previous studies have "concentrated on either psychological factors, or on the perceived attributes of the medium and [...] often used small and specific data sets" (Yates 1996b: 29), Yates (1996b) conducted a large corpus-based analysis in which he compared spoken, written and computer-mediated discourse. The quantitatively most extensive email research, however, was carried out by Helen Petrie in 1999 in a project sponsored by the Microsoft Network (MSN) in which she analysed 38 000 emails[4].

4 In her research, Petrie coins the term "emailism", a term which has to be used with care, for the features that she considers characteristic of email are, in fact, typical of CMC in general and even more distinct in text messages.

Other important experts on computer-mediated communication in the English-speaking world are the Americans Naomi Baron and Susan Herring. Both have published and edited numerous articles on CMC.

In contrast to emails, academic interest in text messages is more recent and there is very little published research on the linguistic features of text messages available. It is not only the factor of privacy which is a problem for researchers in their acquisition of data, but also the ephemeral quality of these messages: "like the oral tradition, text messages are difficult to capture: today's message will not exist tomorrow" (Kasesniemi and Rautiainen 2002: 178). The reason for the remarkable lack of research on text messages in the Anglophone domain, however, is also due to the fact that texting has failed to take off in America. One reason for this phenomenon is given by Thurlow (2003) who explains that in the USA "the internet has tended to be the communication technology of preference" and with it Instant Messaging (IM): "[a]lmost half (46%) of online teens say they most often choose IM over email and text messaging for written conversation with friends. Only a third (33%) say they most often use email to write messages to friends, and about 15% prefer text messaging for written communication" (Lenhart et al. 2005: 14). Moreover, in America talk is cheap. Analyst Jessica Sandin at Baskerville notes that "texting first took off in other parts of the world among cost-conscious teenagers who found that it was cheaper to text than to call [...] but in America, you might as well make a voice call" (*The Economist* 2003). Hence it was only in 2002 that the SMS boom, which was generated by the youth in Europe, reached the USA (Schotzger 2002). Similar to American people, "UK residents do not use SMS as frequently as residents of other countries including the Philippines, Finland and Germany" (Grinter and Eldridge 2001: 220). Accordingly, there have been significantly more studies conducted in Germany[5], for instance, where Höflich, Rössler, Androutsopoulos and Schlobinski have published various articles on text messaging. A major study on mobile communication was conducted by Katz and Aakhus; their results were published in 2002. Further research on CMC

5 In fact, it can even be generally claimed that "more extensive and more detailed studies of language and communication in the Internet have been carried out in the German-language domain than in the Anglophone domain" (Schlobinski 2001b: 2).

can be found in Günther and Wyss (1996) and Schlobinski et al. (1998). Also to the core literature belong Dürscheid and Ziegler (2002) as well as Beutner (2002), Wcingarten (1997), Höflich and Gebhard (2003), and Bittner (2003).

In contrast to the previous studies that have been conducted in the field of CMC, this book analyses private computer-mediated communication, namely private emails, text messages and multimedia messages. No linguistic study so far has compared English written private emails and SMS, and since there has not been any similar study or any close linguistic analysis of English SMS at all, the data cannot be directly backed up by any previous results nor can any direct comparisons to earlier research be made. Moreover, official and private email correspondence differ in various crucial ways: whereas email conversation approximates synchronous interaction in offices, it is of a more asynchronous nature when privately used, because people (and students in the case of this study) do not and are not expected by their peers to sit in front of the computer all day whereas people in offices usually do. Therefore, this book shall, inter alia, be a first step towards the linguistic research of a so far academically neglected field, notably private emails and text messages as well as multimedia messages.

It has been outlined that it is rather difficult to gather an SMS corpus consisting of private text messages. With multimedia messages, however, the situation is even worse. Hardly any research material – and no linguistic study at all – can be found on MMS. Also, there is "not much knowledge on the use of still images in communication" in general (Mäkelä et al. 2000: 549). Since MMS is a rather new phenomenon, penetration rates vary from country to country and it is thus especially difficult to collect a corpus. Hence, most of our knowledge about MMS comes from what Koskinen (2005b) calls "design studies". These are "studies of technology that do not yet exist on the marketplace". Accordingly, in many of these studies

> people have been given access to multimedia phones and an ability to share messages with them. They have received minimal instructions from researchers, who have followed messaging and interviewed people to understand their experience. The justification is that by now, mobile multimedia has been too small a phenomenon to be studied purely from statistics (Koskinen 2005b).

The problems with these design studies, however, are manifold. First, informants knew that they were being studied (cf. Koskinen 2004: 8) which inevitably led to an observer's paradox. Moreover, participants could use the Multimedia Message Service for free (cf. Koskinen 2004: 8) which also had an influence on their use of the multimedia messages. While design studies "tell about first explorations with technology rather than about mature technology" (Koskinen 2004: 9), the analysis of multimedia messages in this book will be a first attempt towards a study that is based on an actual MMS corpus which was gathered from people who did not know that their messages would eventually be used for research purposes. This book will thus contribute to a better understanding of a phenomenon which marks "a whole new era of mobile communications, combining images with sound and text" (Koskinen et al. 2002: 105). In fact, its impact must not be underestimated, for the significance of the evolution from text messaging to multimedia messaging has been compared to the transition from radio to television (cf. Koskinen et al. 2002: 105).

1.3 Methodology: Data Collection and Corpora

The data was collected by means of an electronic survey and a paper-based survey respectively. It was a comprehensive survey which included a questionnaire[6] asking informants about their personal use of email, SMS and MMS as well as gathering demographic information. Moreover, the questionnaire included grids of 160 or more squares onto which informants could copy text messages that they had received. They were further asked to forward three of their private emails and multimedia messages to the email accounts provided.

The survey for the English corpus was conducted between September 2002 and September 2003 at Aberdeen and Bath University and

6 There will not be a close analysis of the questionnaire for most of its questions are irrelevant to this study. Yet, whenever appropriate, some of the results will nevertheless be included.

between July 2005 and January 2006 at various universities in England and Scotland, as well as at Graveney School in Wandsworth/London and at Highfield School in Letchworth (568 students participated, 52% of whom were female, 48% male). The informants and the authors of the collected emails, text and multimedia messages were English native speakers between 17 and 27 years of age.

The same survey was conducted in Switzerland and Germany between July 2005 and March 2006. Those universities and grammar schools which contributed most data to the corpora were the Rheinische Friedrich-Wilhelms Universität in Bonn, the Heinrich Heine Universität in Düsseldorf, the Catholic University in Eichstätt-Ingolstadt as well as the Hardberggymnasium in Bonn and the Max von Laue Gymnasium in Koblenz. In Switzerland, the bulk of data was gathered at the University of Zurich, at the Kantonsschule Ausserschwyz as well as at the Kantonsschule Immensee. All in all, 794 students participated, 54% being female, 46% male.

In none of the surveys was there an observer's paradox; none of the email, text, or multimedia message senders knew that their messages would eventually be analysed for linguistic purposes. However, there was one problem: many emails and text as well as multimedia messages could not be included in the corpus since their owners considered them "too private". In this respect, the sample of messages is not very representative. Nevertheless, the English corpus comprises 342 emails and 983 single and linked text messages with a total of 44 940 and 22 353 words respectively, totalling 67 293 words of text. Not included in the count are any pre- or meta-texts such as message-following adverts or information in the header. The average message length of an email is 131.40 words; the shortest message contains 0[7] and the longest 1192 words. There are only 11 emails that have more than 500 words; one of them was written by a twenty-year-old woman to her twenty-one-year-old partner and is declared a "typed up letter". This implies that emails are not usually of a very great length and probably not as long as letters (which is a hypothesis that lies beyond the scope of this study and cannot

7 In this email, the whole text was included in the subject line, saying "could you make sure [name] reads and responds to email … thanks". There was no text in the body at all.

be proved by means of the available data but would certainly be a worthwhile target for further research). The average message length of single text messages is significantly shorter, namely 19.90 words. This result is similar to Schwitalla's findings: Schwitalla (2002: 47) measures that the average message length is 18.31 words. In terms of characters, there is a total of 87 955 characters (including spaces), which corresponds to 96.97 characters per message. These figures show that even though text messages are restricted in length, they do not make use of all the characters available: the average text message comprises only around half of the 160 characters available. Linked SMS, in contrast, are remarkably longer: they contain 56.62 words and 279.49 characters on average.

There is also a very small English MMS corpus which comprises 27 messages. Since many students still do not possess an MMS-enabled mobile phone or do not use the Multimedia Message Service for economic or other reasons, it was almost impossible to find anyone who has ever used this service in the UK or was in possession of any MMS at all. The collected corpus is thus very small. To still be able to comment on the nature of multimedia messages, the English MMS corpus will be supplemented by German and Swiss-German multimedia messages. Usage tendencies of this new messaging service, such as the text-image proportion or the discursive character of MMS, will be worked out cross-linguistically. There is also a telegram corpus of 73 telegrams, which was collected at flea markets as well as through Ebay, one of the biggest Internet auction platforms.

The German corpus consists of 228 emails, 479 SMS, 36 MMS and 120 telegrams, while the Swiss-German corpus contains 161 emails, 626 SMS and 235 MMS. Both corpora will serve as reference to show and back tendencies which are not necessarily of a linguistic nature, such as the proportion of image and text in MMS, the function of emoticons or – as in Figure 1-1 to 1-3 – the length of emails and text messages across languages. Interestingly, the average length of an email is almost identical throughout all corpora: it consists of 131.40 words in the English corpus, 132.47 words in the German corpus and 131.36 words in the Swiss-German corpus. As for the maximum number of words, the two longest emails are almost equally long in the German and Swiss-German corpora, while the longest English email is somewhat shorter. The result of the minimum number of words, however, is again similar in all corpora.

Figure 1-1: Email Length

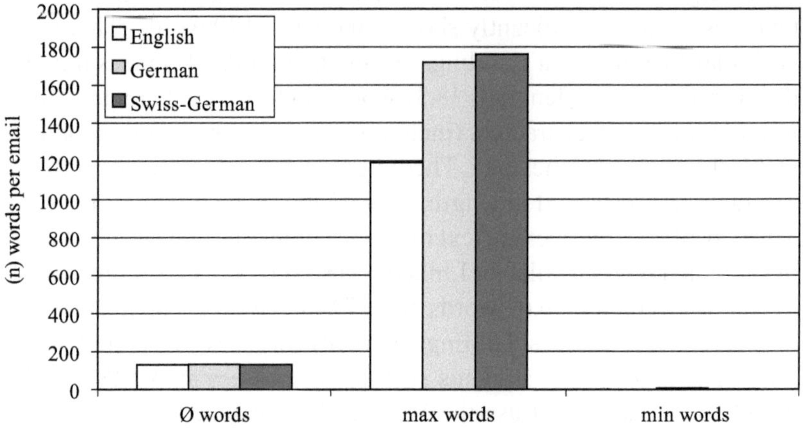

Emails	Ø words	max words	min words
English	131.40	1192	0
German	132.47	1720	6
Swiss-German	131.36	1761	2

Figure 1-2: Single Text Message Length

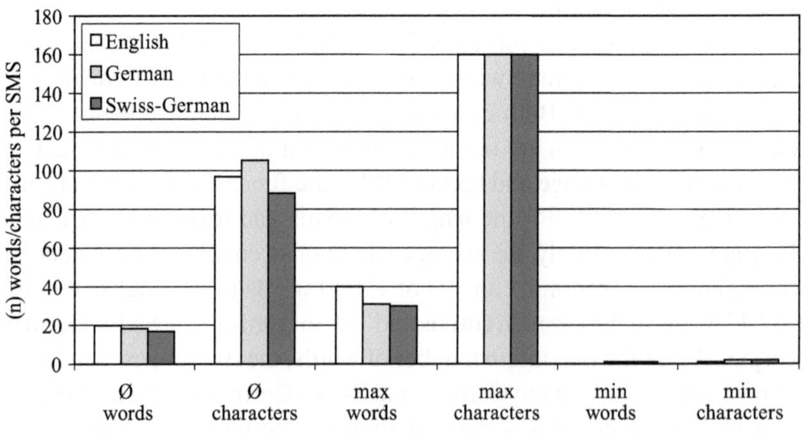

Single SMS	Ø words	Ø chars.	max words	max chars.	min words	min chars.
English	19.90	96.97	40	160	0	1
German	18.35	105.37	31	160	1	2
Swiss-German	16.77	88.11	30	160	1	2

Figure 1-3: Linked Text Message Length

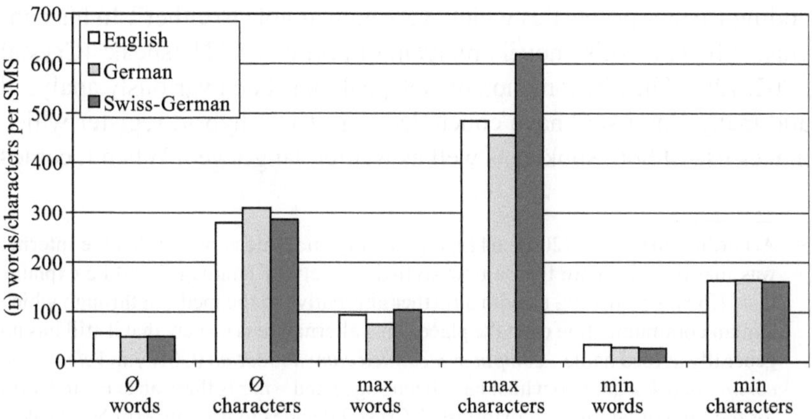

Linked SMS	Ø words	Ø chars.	max words	max chars.	min words	min chars.
English	56.62	279.49	94	485	35	163
German	49.25	309.43	75	456	29	164
Swiss-German	50.72	286.63	104	619	27	161

In terms of text messages, the results are almost equally identical, as Figure 1-2 and Figure 1-3 show. The average number of words in single as well as linked SMS is similar in all corpora and so is the average number of characters. Linked text messages have a greater variance regarding the maximal number of words and characters. Yet, this difference can be neglected as it only concerns peak values. In respect of the minimum number of words and characters, the data of linked text messages is again surprisingly similar.

1.4 Aims and Research Questions

Crystal (2001: 238 f.) finds that "Netspeak is a development of millennial significance" for the reason that a "new medium of linguistic communication does not arrive very often in the history of the race". He explains that "[t]he Internet is a genuine new medium of linguistic communication, taking some of the properties of the two traditional mediums, speech and writing, synthesizing them in a novel way, and adding further properties which were unavailable to either". In his conclusion he calls this "novel amalgam of properties" Netspeak[8] (Crystal 2002: 129). The phenomenon of Netspeak[9] has been variously analysed and many linguists[10] have concluded that it is a hybrid register which makes use of both spoken as well as written language[11]. When it comes

8 According to Crystal (2004: 64 f.), it gradually "became apparent that the Internet was manifesting more than a new stylistic variety of language" and he explains that "[i]t was providing us with a further alternative to the medium through which human communication can take place. This alternative is so new that it still has no generally agreed name – *computer-mediated communication* (CMC) and *electronic communication* are two which have been suggested – nor is there an accepted term for the kind of language it manifests". Crystal decides to use the term Netspeak.

9 The terms Netspeak and computer-mediated communication (CMC) are here used interchangeably.

10 With regard to emails, for instance, Baron (1998a: 155) finds that "[w]hile Social dynamics, Lexicon, and Style distribute themselves neatly on the dichotomous writing/speech spectrum, Format and syntax have mixed profiles". Beutner (2002: 105) speaks of a mixture between orality and literacy while Ferrara et al. (1991: 10) consider interactive written discourse as "a hybrid register that resembles both speech and writing, yet is neither". Similarly, Yates (1996a: 81) explains that "CMC has specific similarities and specific differences compared with speech and writing". This is also true for text messaging. Accordingly, Schlobinski et al. (2001: 6) show that SMS communication exhibits features which are known from the online chat, among them the hybridisation of writing and speech, and Thurlow (2003) concludes that "[l]ike email, and indeed most CMD, text-messages have much the same hybrid quality about them – both in terms of the speech-writing blend but also in terms of old and new linguistic varieties". For further readings, cf. also Meise-Kuhn (1998) or Bittner (2003).

11 Dürscheid (2004: 148) criticises Crystal's term Netspeak and shows that Netspeak as such does not exist. She explains that by means of this term, Crystal tries to construct a unity within a medium which is more heterogeneous than any medium

to text messaging, one can even read of "a new and unique argot" which has been introduced (Vincent 2004: 5)[12]. As a necessary consequence, however, and with the intrusion of oral features into writing, established grammar rules and spelling conventions have been steadily undermined and become unstable. The aim of this book is to characterise the register of emails and text messages in order to find out in what ways, how and why the influence of new media such as email and SMS have changed the English language. There will also be an analysis of multimedia messages as well as a comparison of telegrams with text messages to see how far the telegram-like nature of SMS differs from the traditional telegram style and whether one can truly speak of a telegram renaissance when talking about text messages. Moreover, there are four major research questions that will be addressed in the course of this study, these being the following ones:

- What are the characteristic (conceptually) oral features of email and SMS in terms of economy and efficiency?
- Which are the strategies to substitute the absent paralinguistic cues in CMC?
- Where along the written/spoken continuum can CMC be placed?
- Does CMC lead to the deterioration of the English language or will there be new and generally accepted spelling conventions?

The German and Swiss-German corpora will be used as reference, either to back those statistical features which are too vague in the English corpora or else for comparison reasons. Additionally, there will be an analysis of the Anglophone influence on German and Swiss-German

has ever been before. In fact, there are so many different kinds of text types which are nowadays written online that one can no longer make any generalising statements in regard of the occurrence of certain linguistic features and Schlobinski (2001b) concludes that "[t]he handy term *Netspeak* seems […] to be non-felicitous from linguistic and communication theory perspectives, especially when it is brought into context with a concept of variety which is not defined further". He explains that linguistic elements from various worlds of discourse have led to a specific mixture of style so that one can neither speak of an Internet language nor of an extralinguistic variety (Schlobinski 2000 in Dürscheid 2004: 149).

12 Cf. also Taylor and Vincent (2005: 76–84).

messages to see how far and in what way the German language has become Anglicised. Beutner (2002: 87), for instance, claims that the English influence in CMC in general is conspicuous despite the fact that she cannot find many messages in her corpus with a noticeably larger use of English vocabulary. Similarly, Weingarten (2002) points out that the application conditions of the computer have not only led to a Latinisation, but also to an Anglification of the written language system. He further asserts that with the globalisation of communication, the English language has become a lingua franca (Weingarten 2002 in Gerber et al. 2002: 167). An analysis of Anglicisms in the German and Swiss-German corpora will show whether the use of English vocabulary is indeed as prevalent as commonly claimed, and whether there are any usage differences in emails and text messages.

The study combines quantitative with qualitative methods and restricts itself to the asynchronous media, i.e. email, SMS and MMS. The aim is to see in what ways the written language has changed and become more like the spoken. Moreover, only emails via the Internet and SMS/MMS via mobile phones are included because WAP[13]-emails as well as Internet-SMS/MMS underlie different conditions which inevitably influence the composition of the messages. The computer keyboard, for instance, makes typing much more comfortable than a mobile phone keypad and thus reduces the necessity to shorten words. And since Internet-based text messages often include adverts or information of the Internet service used, the message length of Internet-SMS may be significantly reduced.

13 WAP, short for Wireless Application Protocol, is a "protocol for providing wireless devices with secure access to text-based information, including Web pages, chat, and e-mail" (Jansen 2002: 405).

1.5 The Register of Emails and Text Messages

Certain technical means are necessary to send and receive emails and SMS or any computer-mediated messages, and they are thus classified as tertiary media[14]. Both constitute a form of non-auditory and non-visual communication. For this reason, they lack the paralinguistic cues. This is why email and SMS are often considered cold and impersonal, which is one of the predominant factors that restrict the type of messages that are accepted for these media. According to the students interviewed, email and SMS are mainly used to send greetings, make arrangements, share problems, gossip, or else are written out of boredom, but are not chosen for negative content such as appointment cancellations, the breaking off of a relationship or any kind of bad news, which should be, so the general consensus, communicated orally or via a handwritten letter. One informant says that she would never tell anything serious such as bad or very personal information and concludes that her emails and text messages only include "really happy things", while another informant advises "never dump a boyfriend by email or SMS". Similarly, Crystal (2001: 126) states that "there is a widespread feeling that letters are better than e-mails for expressing negative content [...] and that telephone or face-to-face conversation is also better in such cases, where the full range of vocal nuance is needed to do justice to the meaning".

Though SMS is a mobile communication medium, Höflich and Rössler's (2000: 10) survey shows that the majority of text messages are written at home. Yet, the mobile phone also prompts interaction in situations where people would not have communicated before. Ling and Yttri (2002: 139) observe that "one of the impacts of mobile telephony is the opportunity for nuanced instrumental coordination". Whereas people could not be contacted on the road before the advent of the mobile phone, they are now

14 According to technical criteria, one generally differentiates between primary, secondary and tertiary media. One speaks of primary media if no technical means are required to communicate, as in the case of face-to-face communication. Secondary media, in contrast, require a technical means on the side of production but not on the side of reception (e.g. print media), while tertiary media demand technical means on both sides (e.g. radio, television, email, etc.). (Dürscheid 2005)

reachable almost anywhere, which encourages spontaneity and ad-hoc coordination: "With the use of mobile communication systems, one need not take an agreement to meet at a specific time and place as immutable. Rather, those meetings have the ability to adjust the agreement as the need arises." (Ling and Yttri 2002: 139) Such pre-face-to-face communication can serve different purposes like modifying an appointment (2), assuring an arrangement (3), or expressing excited anticipation (4).

(2) Hey my mum went to work without leaving me money for the train.Just text me and Ill meet you later
(3) I WILL BE OVER SOON
(4) V EXCITED BOUT COMIN HOME!C U IN A HR.TAKE CARE LX

Besides pre-face-to-face communication, this new form of interaction also provides post-face-to-face communication as in example (5), for instance, in which the sender comments on the meeting. Moreover, text messages can also be used as a means of meta-communication, coordinating further communication via another medium by announcing a telephone call (6) or an email (7).

(5) Hi [Name]! Thank u so much 4coming2meet me,twas a lovely surprise!Im on the train Aberdeen+planning2head home 2nite.Looking4wrd 2seeing u on sat![name]x
(6) Going to sleep now will can [= call] you in morn.when did we say for call?hope you have a good and safe journey. Goodnight. Xxxxxxxxxxxx (my comment)
(7) Email kept failing last nite. [Name] is still here. Willk email 2nite

Often, text messages are of a mere phatic nature "where the point is not so much the message but the gesture of getting in touch" (Haddon 2000: 5) as illustrated in example (8). The same applies for emails, but email is not a preferred medium for pre- or post-face-to-face communication because it is not mobile and therefore not as immediate as SMS.

(8) How are you doing [Name]? I'm better now... On the rough ride hehe. How is Uni?

Emails, in contrast to text messages, are more substantial in content and longer but not suitable when a quick response is needed. Not being restricted in length, they are also more narrative than text messages and can be used to update family or friends on the latest news, to talk about exciting (holiday) experiences or anything which is of interest but not too urgent to be communicated via a (near-)synchronous medium.

1.6 Emails and Text Messages along the Written/Spoken Continuum: A First Approximation

Several studies have suggested that the relationship between writing and speech should not be considered a dichotomy but a continuum (cf. Biber 1988; Koch and Oesterreicher 1985, 1994; Haase et al. 1997), while other studies debate this concept, arguing that "the idea of one continuum or several continua is too simplistic a view of the relationship" (Barton 1994: 88). Barton (1994: 88–90) explains that "we find written and spoken so intertwined that good writing contains oral strategies" whereas "spoken language is an important context for most literacy events". In other words, written and spoken language "are not actually separable in real life" and especially in the forms of language associated with new technologies, the written aspect cannot be readily disentangled from the spoken aspect (Barton 1994: 90). This kind of language is, in fact, a hybrid; it mixes the spoken features with written ones and thus does not belong to writing or speech.

Chafe and Danielewitz (1987: 92) describe oral language as "more innovative" and written language as "more conservative". The use of new words and new senses of old words as well as contractions are among the examples of innovative spoken vocabulary. Moreover, spoken language contains relatively brief intonation units "because speakers can focus their consciousness on only a limited amount of material at one time" (Chafe and Danielewitz 1987: 97) and "there is a strong tendency for casual speakers to produce simple sequences of coordinated clauses" that are often connected with the word *and* (Chafe and Danielewitz 1987: 103).

Koch and Oesterreicher (1994: 587) go a step further. They distinguish between medium and conception and state that the terms written/spoken are often used for two different concepts: first, they refer to the medium of realisation where oral would be equalled to phonic and written to graphic. Second, they refer to the ductus, the modality of the utterance as well as the linguistic varieties used, i.e. the conception which coins the utterance. They further point out that a private letter is closer to conceptual orality despite its realisation in the graphic medium (written mode), a statement which also applies to emails and text messages. With regard to the medium, both emails and text messages are written; conception-wise, however, they combine

> qualities which are typically associated with face-to-face interactions, e.g. immediacy and informality of style, transience of message, reduced planning and editing, rapid (or immediate) feedback with properties of written language, e.g. lack of visual and paralinguistic cues, physical absence of the addressee, written mode of delivery, etc. (Georgakopoulou forthcoming).

To establish where (if at all) email and SMS are integrated along the written/spoken continuum, it is helpful to consider Biber's theory of the six dimensions of register variation[15]. In his first textual dimension in speech and writing, labelled "Informational versus Involved Production", Biber (1988: 115) compares "discourse with interactional, affective, involved purposes with highly informational purposes, which are carefully crafted and highly edited". He then puts the different genres in relation to each other and finds that "face-to-face and telephone conversations have very high means scores on Dimension 1", while personal letters only have "moderately high scores" and are thus clearly distinguished from face-to-face or telephone conversation. Genres with high means scores are characterised by frequent occurrences of private verbs, *that*-deletions, contractions, present tense verbs, second person pronouns, *do* as a pro-verb, analytic negation, demonstrative pronouns, general emphatics, first person pronouns, the pronoun *it*, *be* as a main

15 The six dimensions are "Informational versus Involved Production", "Narrative versus Non-Narrative Concerns", "Explicit versus Situation-Dependent Reference", "Overt Expression of Persuasion", "Abstract versus Non-Abstract Information" and "Online Informational Elaboration" (Biber 1988).

verb, causative subordination, discourse particles, indefinite pronouns, general hedges, amplifiers, sentence relatives, *wh*-questions, possibility modals, non-phrasal coordination, *wh*-clauses, and final prepositions. At the same time, these genres show infrequent occurrences of nouns, long words, prepositions, varied vocabulary, and attributive adjectives. For genres with low means scores, the opposite is true (Biber 1988: 102–135). Figure 1-4 illustrates the relations among Biber's analysed genres with respect to the informational versus involved production of Dimension 1. It further summarises the factorial structure of this dimension and lists the features according to their positive and negative weights. Following Biber (1988: 101),

> the features with positive and negative weights on a factor have a special relationship to one another: the features with positive weights co-occur in texts; the features with negative weights co-occur in texts; and these two groups of features occur in a largely complementary distribution. That is, when a text has several occurrences of the features with negative weights it will likely have few of the features with positive weights, and vice versa.

The difference between face-to-face and telephone conversations and personal letters is mainly due to the factor of "time constraints" (Biber 1988: 108). In contrast to personal letters, oral conversation is "produced under real-time conditions" and thus "constrained in its lexical precision and informational density" (Biber 1988: 108). The question now is where email and SMS are to be arranged along this dimension. Basically, they are both comparable to personal letters, as both of them are characterised by an involved style while being interactive and affective. Yet, whereas letters are, in any case, a means of asynchronous communication and are thus produced more carefully and edited[16], emails and SMS can approximate synchronous discourse; they are written online, underlie real-time conditions, and tend not to be reread nor edited (cf. Crystal 2001 and Baron 2000). In this point, email and SMS are highly

16 "In a paper document, it is absolutely essential to make everything completely clear and unambiguous because your audience may not have the chance to ask for clarification. With email documents, your recipient can ask questions immediately. Email thus tends, like conversational speech, to be sloppier than communication on paper." (Sherwood 2001)

Figure 1-4: Means Scores of Dimension 1 and its Factorial Structure

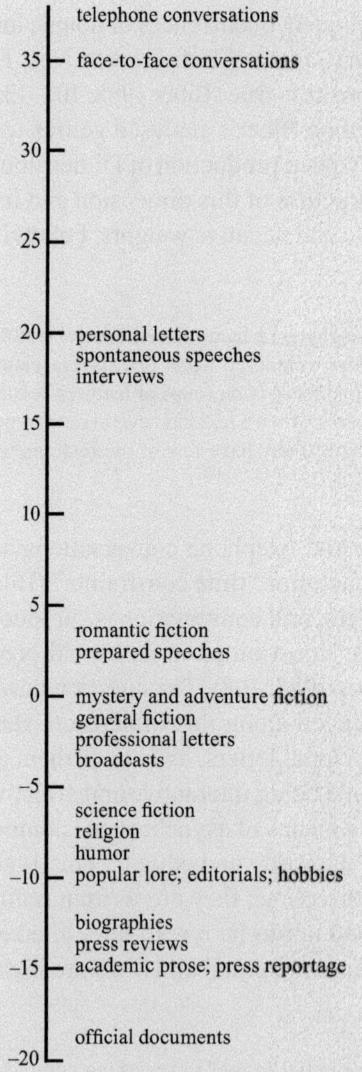

FACTOR 1

private verbs	0.96
THAT deletions	0.91
contractions	0.90
present tense verbs	0.86
2nd person pronouns	0.86
DO as pro-verb	0.82
analytic negation	0.78
demonstrative pronouns	0.76
general emphatics	0.74
1st person pronouns	0.74
pronoun IT	0.71
BE as main verb	0.71
causative subordination	0.66
discourse particles	0.66
indefinite pronouns	0.62
general hedges	0.58
amplifiers	0.56
sentence relatives	0.55
WH questions	0.52
possibility modals	0.50
non-phrasal coordination	0.48
WH clauses	0.47
final prepositions	0.43
(adverbs	0.42)
(conditional subordination	0.32)
nouns	–0.80
word length	–0.58
prepositions	–0.54
type/token ration	–0.54
attributive adjs.	–0.47
(place adverbials	–0.42)
(past participial WHIZ deletions	–0.38)
(present participial WHIZ deletions	–0.32)

(Biber 1988: 102, 128)

comparable to face-to-face or telephone conversation and differ from personal letters. Accordingly, email and SMS can be arranged somewhere between personal letters and face-to-face and telephone conversation. Their language is a mixture of the written and spoken language and has "the spontaneity and informality of speech, though it is 'written' and thus carries with it the written language" (Moran and Hawisher 1998: 94).

The hybrid character of electronic language also comes to light when considering how this language is named: already in 1986, electronic language was described as "talking in writing" or "writing letters which are mailed over the telephone" (Spitzer 1986 quoted from Collot and Belmore 1996: 14). With the advent of the mobile phone, electronic language is further termed text-talk or text-speak, both being paradoxical terms that express the hybridism of this register: while 'text' is generally associated with written substance[17], 'talk' and 'speak' indicate the conceptual orality of these messages.

17 This definition is true in the context of English text linguistics where the term 'text' refers to written text only. Yet, it does not constitute a universal definition and so, the word 'Text' in German text linguistics, for instance, may include written as well spoken 'Texte' (cf. Stein 2001).

2 Email

2.1 Definition

Email (electronic mail) is electronically transmitted mail via computer. Depending on its size, which is weighed in bytes, the transmission of data takes places within a few seconds or minutes, whereby the loss of data is less than 0.05% (Kümmel 1997: 6). A major feature of email is its asynchrony: users do not have to be online simultaneously, as an email message can be composed, sent and read at any time from any Internet-connected computer. Email is still stationary but there have been first steps towards a mobile email service: some email services notify the receiver about an incoming email via SMS and it is further possible to send and receive emails via WAP-enabled mobile phones. The majority of users, however, still send their messages via Internet; WAP emails have not yet caught on.

As early as 1988, Schlobinski et al. (1988: 28) described email as the most widely used Internet service and the results of the 2003 survey by the Pew Internet and American Life Project support this observation: 52% of American adults went online in June 2003 and 49% used the Internet to send emails. Yet with the rapid change and progress of technology, users' habits tend to change equally fast. Accordingly, a creeping shift from email towards another highly popular form of communication has been noticed within only a few years: while email has been the most popular online application for many years – "a communications tool with a popularity and a 'stickiness' that keeps users of all ages coming back frequently" – newest surveys indicate that "email is losing some of its appeal" as a growing number of especially young users "express a preference for instant messaging [IM]" (Lenhart et al. 2005: 14). Nevertheless, "[t]he presence of email in teens' lives has persisted, and the number that uses email [92%] continues to surpass those who use IM [84%]. But when asked about which modes of communication they use most often when communicating with friends, online teens consistently choose IM over email in a wide array of contexts" (Lenhart et al. 2005: ii).[1]

2.2 Historical Background: Email as By-Product of the US Military Defence System

Email is a relatively new form of communication that has been available to the general public only since the mid-1990s. Yet, its invention goes back to the year 1971 when the first email message was transmitted in the United States; the first idea of email, however, came up more than a hundred years ago:

> In an 1863 manuscript, "Paris in the 20th century", found and published for the first time in 1994, Jules Vernes wrote, "Photo-telegraphy allowed any writing, signature or illustration to be sent far away, and any contract to be signed at a distance [20,000 km]. Every house was wired." (Negroponte 1995: 190)

---

1 In Europe, teenagers are also using email less and less. While it was rather simple to gather an email corpus in the years 2002 and 2003, it proved significantly more difficult in the years 2005 and 2006. Many informants explained that they no longer used emails for private peer-to-peer correspondence, but rather as a tool of conversation reserved for communication with adults such as teachers. Similarly, Longmate and Barber (2002: 85) conclude that email is not only infrequently checked and used in comparison with SMS, but that it is also rarely used between students. Bittner (2003: 147) points out that email is originally a means of distance communication while Longmate and Barber (2002: 78–84) summarise that "[e]mail appears to support longer distance relationships than SMS" and is more used for information purposes whereas SMS "is often more about contact then [sic] content". They further assert that "in comparison with SMS email supports weak ties or non-close relations" while "[r]elationships supported by SMS have their origins in face-to-face meetings" (Ling and Yttri 1999 quoted from Longmate and Barber 2002: 85). These quotes clearly prove that email is indeed used infrequently for private communication among teenagers and far less regularly than SMS. Yet, it lies beyond the scope of this study to provide more detailed data on this trend. It would, however, be interesting to follow it further, especially as it promises imminent changes with a considerable impact in various respects.

And Baron (1998a: 140) notes that

> [t]he family tree of email predates modern computing by more than a century. The earliest teletechnologies – Samuel Morris' telegraph in 1838 and Alexander Graham Bell's telephone in 1876 – made it possible to send messages at distances in near-real time. The telex, developed in the 1950s, joined copy machine technology with telephone transmission.

Email is a by-product of the US Military Defence System. It "grew out of research projects during the height of the Cold War aimed at developing a decentralised network of computers that could be used to transmit information across the US in case of nuclear attack" (Baron 2000: 223–225). Therefore, a wide area network, the so-called Advanced Research Project Agency Network (ARPANET), which was run by the United States Department of Defence, was built up by 1968; its main aim was to transfer data from one computer to another in a local network. Those electronic message programs, however, were not comparable to today's email protocols, as they only worked locally: they were "designed to allow exchange of messages between users who shared the same machine" (Campell 1998). It was only in 1971, when computer engineer Ray Tomlinson – working for Bolt Beranek and Newman, the company that built the ARPANET – was experimenting with one of the electronic message programs and finally managed to send messages to computers others than linked to the local network. He used the symbol @ "to distinguish between messages addressed to mailboxes in the local machine and messages that were headed out onto the network" (Campell 1998). "The @ sign seemed to make sense," he recalled, "I used the @ sign to indicate that the user was 'at' some other host rather than being local" (Campell 1998).

Even though the first message was sent in 1971, it was not until 20 years later that email could be publicly used, since the United States Department of Defence kept this new technology secret until ARPANET was decommissioned in 1990 (Seasonal Web 1999). Today, email has established itself as one of the most wide-spread modes of communication: "[i]n only three decades, email has grown from a government-initiated, academically-implemented system for sharing information into an international alternative to long distance phone calls, interoffice memos, and face-to-face encounters" (Baron 1998a: 134). In fact, "there are now more e-mail addresses than telephone numbers in the world" (Jansen 2002: 156).

2.3 Structure: Header – Body – Signature

The structure of an email message follows a fixed pattern of functional elements that is "dictated by the mailer software [and] has become increasingly standardised over the past twenty years" (Crystal 2001: 94). Each email contains a header, body and an optional signature. The header, which can be compared to the letterhead in non-electronic mails, consists of four parts, namely the sender's email address, the recipient's email address, the time and date, and the subject. Thereby, the date and sender's line are always filled in by the email program, which makes it difficult[2] to write an anonymous email.

An email message can have one or several recipients. If an email is sent to more than one person, the writer has the option to send his[3] message as carbon copy (by filling in the CC line) or as a blind carbon copy (by filling in the BCC line). The latter has the advantage that the recipient does not see the email address(es) of the other receiver(s), which, of course, also means that "the distinction between one to one and one to many is blurred" (Baron 1998a: 146), for the addressee never knows if he is the only one to get the message or whether it is addressed to further recipients.

The subject in the header can be compared to the reference line in (business) letters. If not left blank, it usually contains a brief description or the keyword of the message and can thus be used as a contextualisation

2 There are still ways to send anonymous emails: one can, for instance, choose a nonsense email address such as anonymous@hotmail.com so that the receiver has no idea who has written the email or, as this is the case in spam and junk mails, there are computer programs that change the name and address of the sender in such a way that the identity of the sender is blurred.

3 For the sake of simplicity, I will not distinguish between female and male forms but use the generic masculine instead. Yet, it shall be emphasised that whenever I refer to the male form, the female sex is included, too.

cue to make intertextual references (Günther and Wyss 1996: 65). It is useful especially if the threads get longer and helps to contextualise the message by maintaining the reference object.

An email can also contain attached documents, e.g. graphics, photographs, sounds, programs or video clips, in fact any kind of data which can be transferred in form of binary codes. Such documents are known as attachments and are comparable to the enclosure in (business) letter mails. Depending on whether there is an attachment added to the message, the header may also include a space that contains the attachment (usually indicated by a paper-clip), while an extra space is provided for the degree of priority: the sender can choose between low, normal and high priority. The degree of priority shows whether a message is urgent, but it has nothing to do with the speed of transmission which will, in any case, remain the same.

As for the body, it can consist of pure writing, but can also include graphics, photographs or sounds. At the end of the body, there may be a signature[4] which is stored in an extra data file and automatically added by the email program. This signature is optional and usually contains further information of the addresser such as phone- or fax-number, or postal address, but it can also be made up of jokes or quotations.

2.4 Rules and Conventions: Netiquette

The neologism netiquette is a blend of 'net' and 'etiquette'. It provides rules of conduct when communicating through the Internet. The first netiquette, titled *Toward an Ethics and Etiquette for Electronic Mail*, was composed by Shapiro and Anderson and published in 1985. Since then, various netiquette guidelines have been written and made available either via the Internet or in the form of book(let)s. Many of these

4 Since the signature contains further information about the addresser, it can be equalled to the letterhead in (business) letter mails. The only difference is that the signature is added after the body whereas letterheads, as the term indicates, always precede the body.

guidelines, however, are directed towards interaction in business, academia, Usenet or mailing lists and thus do not necessarily apply to private one-to-one or one-to-many email correspondence. Also it is doubtful whether people who use email for private purposes ever make themselves familiar with these rules and conventions.

Still, netiquette constitutes an important guideline, as emailing is a new mode of communication with which the public has been familiar for only about two decades whereas "people have had about 50,000 years' experience in the use of speech and gestures, 5,000 years' experience in writing, and about 100 years' use of the telephone" (Shapiro and Anderson 1985). They know the rules and regulations of these traditional modes of communication, but are now faced with the problem that these rules do not correspond to email: "you do not write business letters as electronic messages [...] and yet electronic messages are not printed telephone conversation either" (Shapiro and Anderson 1985). In other words, there is a lack of a historically grown etiquette that is instinctive to the user (Freyermuth 2002: 9). Hence, problems arise quite easily and misunderstandings are not rare. In order to prevent these complications, a new set of rules is needed which tells the user "how to be a constructive, courteous sender and receiver of electronic messages" (Shapiro and Anderson 1985). Many of the guidelines indirectly postulate Grice's four Maxims of conversation when advising the user:

1) not to be a novelist[5] and to be selective in broadcasts for information[6] (*Maxim of Quantity*);
2) to assume that any sent message is permanent[7] and can be seen by other people, i.e. to never send anything that one would not want to see in tomorrow's newspaper[8] (*Maxim of Quality*);
3) to create single-subject messages whenever possible[9] and to be concise and to the point[10] (*Maxim of Relevance*);

5 Cf. I Will Follow Services™ (2003).
6 Cf. Shapiro and Anderson (1985).
7 Cf. Shapiro and Anderson (1985).
8 Cf. Linfield College (2003).
9 Cf. Shapiro and Anderson (1985).
10 Cf. IEEE (2002).

4) to avoid responding while emotional or else, label emotional outbursts (e.g. by using emoticons) to reduce shock, surprise or disappointment in the reader that are normally avoided by other social conventions in face-to-face interactive conversation[11] (*Maxim of Manner*).

Yet, these rules are not binding and not necessarily obeyed. In other words, despite the numerous stylistic guidelines and netiquette regulations, there are still no generally valid conventions – not only in terms of the conversational maxims, but also in terms of pragmatics. On the one hand, this is due to the fact that emailing is a very novel way of communication and still in the process of developing its own conventions; on the other hand, it makes use of the features of unplanned discourse for which, on a linguistic basis, no explicit rules – neither descriptive nor prescriptive ones – have been established. Milroy (1985: 146) notes that "descriptive grammars, our best source of information on the pattern underlying English utterances, reflect the structure of planned rather than unplanned discourse". Since email is a matter of unplanned discourse, "characterised as containing 'errors' or as 'unstructured' in relation to the norms of planned discourse" (Milroy 1985: 146), mistakes in spelling, punctuation and syntax are largely tolerated. This makes email an especially easy-going medium whose informality also encourages those people to communicate in writing who do not normally write letters and who are therefore not too familiar with the traditional conventions of letter writing. Moreover, "norms of spelling and punctuation in CMC may be increasingly in flux" (Cho: forthcoming) for the reason that "a growing number of people with low levels of English language skills gain Internet access" (Cumming 1995 quoted from Cho: forthcoming) and so, even though the existence of netiquette dates back about ten years, a common set of clear conventions has neither been established nor is it likely that any of them will gain general acceptance in the near future. What is more: "[s]ince we construct (and send) email in social isolation, and since we see the medium as ephemeral, we don't feel particularly constrained by the social conventions that govern face-to-face exchange or written communication" (Baron 2000: 235).

11 Cf. Shapiro and Anderson (1985).

2.5 Features of the Involved Written but Conceptually Oral Language

2.5.1 Salutation and Farewell

Emails, in contrast to ordinary letters, do not necessarily include a salutation or farewell. As there are no clearly established stylistic conventions, greeting and leave-taking formulas may be omitted or appear in various unconventional ways. This, however, also unsettles email users so that many of them "indicate they are unsure of the salutation and closing format to employ" (Horowitz and Barchilon 1994: 209): they clearly distance themselves from the simple and traditional formulas such as *dear* <recipient's name> or *love* <sender's name> and employ a great variety of different and new ways of salutations and farewells. Farewells tend to be formulated more carefully and are rather lengthy at times. In contrast to salutations, which are more traditional, farewells are hardly designed along the same lines: it is as if no restrictions are placed on creativity, as examples (9) to (11) illustrate.

(9) lots and lots of love, and thank you for all your support recently. love [name]
(10) hugs/kisses/mad jumping up-and-down-up-and-down/more hugs [name] (tired [name])
(11) You take care [addressee's name] and do email me, will check asap! Love you lots [name]

These findings support Kiesler and Sproull's observation that "email senders tend to use more words for the closing than for the salutation" (Kiesler and Sproull 1986 quoted from Horowitz and Barchilon 1994: 210). It seems as if the sender of a message tried to compensate the anonymity and aloofness of the medium with extended leave-taking formulas. Consequently, salutation formulas contain 1.25 words on average, while farewell formulas consist of an average of 3.92 words.

As there are so many different salutation and farewell formulas, they are best summarised by means of categories. Thereby, the plurality may get lost, but the basic elements which are employed can most clearly be worked out this way. Table 2-1 and Table 2-2 summarise the frequency

Table 2-1: Salutations in Emails

Salutation categories	number	%
<first-/nickname>	13	4%
dear	7	2%
hello	25	7%
hey	71	21%
heya/hiya	19	6%
hi	94	27%
others	9	3%
none	104	30%
total	342	100%

Table 2-2: Farewells in Emails

Farewell categories	number	%
<first-/nickname / initials>	61	18%
all the best	7	2%
bye	5	1%
(big / lots of) hugs	12	4%
cheers	7	2%
(all my / lots of) love / love you	87	25%
regards	6	2%
see you <time adverbial>	14	4%
talk/speak <time adverbial>	16	5%
take care	29	8%
thanks	10	3%
your/s <first-/nickname>	3	1%
others	34	10%
none	51	15%
total	342	100%

of salutation and farewell categories in the analysed corpus. In contrast to Cho (forthcoming) and Herring's (1996: 87) observation, the formulas in this corpus are not as often neglected as in their corpora: salutations are omitted in 30% of all the messages while only 15% of the farewells are left out. Cho's formulas, however, are omitted in about one third of the messages, whereas Herring reports that salutations are absent in 87% and farewells in 98% of her analysed messages. This remarkable difference might be due to the fact that both Cho and Herring's sample of messages are not private but business emails and messages from Internet mailing lists respectively. Still, the result is surprising, as one would assume that formal emails, unlike personal informal ones, rather adapt themselves to the traditional rules of letter writing and thus contain greeting and leave-taking formulas, but this assumption has clearly proved wrong.

Interestingly, there are only seven instances (2%) of the traditional *dear*-greeting while most of the salutations (61%) consist of so-called *h*-salutations, i.e. variants of the interjection *hello* such as *hey*, *hi* or *heya/hiya*. This result presents a challenge to Crystal's (2001: 101) observation who finds that the salutation *dear* was "by far the most frequent individual greeting formula", but backs up Yongyan's (2000: 31) finding that "correspondents rarely adopt a formal formula of the type 'Dear […]'" Crystal (2001: 102) further states that "farewells display fewer possibilities for variation [than salutations]" which again does not apply to the analysed corpus of this study: due to the various different combinations of closing formulas such as *take care*, *love*, *hugs*, etc., there is a far greater range of farewells than salutations. Still, 18% of the emails conclude with the sender's name or initials only, 25% with (variants of) the traditional *love* formula, and 8% with the phrase *take care*. As Cho (forthcoming) rightly states, such formulas clearly serve phatic purposes and as such express gratitude, affection, wishes, expectations, requests for further contact, etc., allowing the sender to attribute the email with an individual touch. In fact, whenever extended farewells are employed, they are unique and not used in the same way by any other person, as the above examples (9–11) clearly illustrate.

In contrast to a one or two turn email conversation, email dialogues are bound to lack greeting- and leave-taking formulas, but this is the topic of Section 6.1, Email and Text Message Dialogues. So far, it can be concluded that farewells, more than salutations, are an important feature of discourse in private emails and largely make up for the anonymity and coldness often felt when communicating via tertiary media.

2.5.2 Spelling

2.5.2.1 Misspellings, Typos and Transfer Errors

Very typical of the language of email and computer-mediated communication in general is the fact that less attention is paid to spelling and typos. This is largely due to the speed with which a message is composed: one tries to write as quickly as one thinks while bearing in mind that everything that is keyed in can be deleted, modified or corrected at any time. Hence, the computer dramatically changes people's writing habits: the sender of a message no longer cares about carefully composed and beautifully adorned sentences, but spontaneously writes down what comes to his mind as if he were speaking. Johnson (1997: 143f.) describes his experiences with the word processor as follows:

> I began with my familiar start-and-stop routine, dutifully thinking up the sentence before typing it out, but it soon became clear that the word processor eliminated the penalty that revisions normally exacted. If the phrasing wasn't quite right, you could rearrange words with a few quick mouse gestures, and the magical 'delete' key was always a split second away.

Thereby, Johnson (1997: 144) rightly recognises, "the thinking and typing processes beg[i]n to overlap" and so, there is no time for immediate correction or rereading sentences as this would not only interrupt the typing but also the thinking process: "[a] phrase," Johnson continues, "would come into my head – a sentence fragment, an opening clause, a parenthetical remark – and before I had time to mull it over, the words would be up on the screen".

Efficiency and economy are the crucial factors when composing a message almost as fast as one thinks. To this end, the language of email "relies principally on expression which is QWERTY-driven (i.e. whatever is possible with the computer keyboard) and economic (i.e. saving a keystroke wherever possible)" (Thurlow 2001: 288). In such circumstances, spell- and grammar-check options are hardly useful and that is why "many users of word-processing and email programs disable these correct-as-you-go features, preferring self-editing, post-editing, or, particularly in the case of email, no editing at all" (Baron 2000: 242). A further important aspect which has to be taken into consideration is that "[p]eople sending email often reveal an editorial nonchalance, reflecting the casual tone of the medium and a psychological mind-set (however mistaken) that email, like telephone, is ephemeral" (Baron 2000: 241).

Since the computer influences the composer's lax attitude towards spelling and encourages abbreviated or economised word forms, it is not always obvious whether a particular misspelling is a mistake or typo at all, for it could as well be an ad-hoc abbreviation and thus an intentionally chosen way of spelling. Hence, it is unclear whether the word *snding* in example (12) is a typo or, as the word *spk* in examples (13) and (14), a matter of consonant spelling[12].

(12) sorry about *snding* texts, i was only told off by [name] yesterday that you have to pay loads to retrieve them. (My emphasis)
(13) dunno if i am dancing tonight so will *spk* to u later. (My emphasis)
(14) take care, and *spk* soon. (My emphasis)

It may be argued that the word *snding* does not constitute an instance of consistent consonant spelling for it still contains the vowel /i/ in its *ing*-form. Yet, it could well be that consonant spelling simply affects the stem of the word and not its ending, but there are not enough examples in the analysed corpus to prove this hypothesis, so further research on this topic would be necessary[13]. What can be said, however, is that the word *snding* is a borderline case: it could be an instance of consonant spelling as well as a typo. Similarly, it is not evident how far the words *no* meaning 'know' and *mos* meaning 'most' in examples (15) and (16) are intentionally spelt this way:

12 Consonant spelling is a kind of spelling which omits the vowels in a word. Since consonants are the meaning-bearing letters, the word can nevertheless be understood, especially when embedded in its context (for further information, see Section 2.5.2.3, Non-Conventional Spelling).

13 Cf. also footnote 12 in Section 3.5.2.3, Non-Conventional Spelling.

(15) Its funny going into second year and staying in student accomadation where you have to get to *no* people again. (My emphasis)

(16) So look forward to seeing *mos* of yous back at uni. (My emphasis)

Both words are pseudo-phonetically spelt as they are pronounced: the former lacks the consonant /k/ as well as the consonant /w/ and the latter drops the consonant /t/, still, both could as well be instances of misspellings or typos. The same is true for the word *completly* in example (17) in which the omission of the vowel /e/ either constitutes an economic measure or an actual mistake:

(17) Do you want me to *completly* remove everything and put plain vanilla Win98 on Plus Office 2000 etc..? (My emphasis)

Transfer errors, in contrast to misspellings and typos, are obvious mistakes that result from wrongly transferred data. There are two kinds of transfer errors (also known as Hypertext Markup Language (HTML) errors) in the analysed corpus: once, the apostrophe is wrongly reproduced as *â?T* (18) and in another case, some of the spaces are erroneously transferred as * * (19).

(18) I*â?T*m just writing to say bye. Leaving for London tomorrow (Friday) at 9pm and I*â?T*m sorry I haven*â?T*t been able to meet or call so many of you (My emphasis)

(19) Worrying* *may make things worse so please try not to* *stress out about things […] im not sure i have the right number.* *By the way, if it is the middle of the night here, just gimme a txt and im* *awake and listening ok? (My emphasis)

Nowadays, however, transfer errors are rare for most email programs support HTML emails, also known as graphic rich emails. This was certainly not the case in the early years of emailing when HTML, as opposed to plain text, came into use and was only supported by a few email programs. This resulted in transfer errors of the kind illustrated above.

2.5.2.2 Use of the Upper and Lower Cases

One of the effects of linguistic economy in computer-mediated communication is the tendency to reduce the use of capital letters to a minimum or omit them at all and compose the whole text in lower case. Another possibility, of course, would be consistent capitalisation, but as capital letters are interpreted as shouting in computer-communicated communication[14], this is not a viable alternative. Hence, there are just three emails in the analysed corpus which are wholly capitalised. One of them, however, merely consists of one word (20), whereas the second one is a short thank you for an invitation (21).

(20) MOOO
(21) THANKS FOR THE INVITE!! REGARDS, [NAME]
(22) AW, I DIDN'T THINK YOU GOT THAT EMAIL!!! DID YOU GET MY OTHER ONE? WHEN YOU GETTIN SOME CRDT? IS THIS JUST GONNA BE BOUT SEXUAL STUFF OR DYA WANNA TALK TOO? MISSED MY LECTURE THIS MORNING, BUT MANAGED THIS AFTERNOON!!! ME AND [NAME] AREN'T GONNA BE TALKING FOR A WHILE, SO I'M ALL YOURS BABY!!! WRITE BACK SOONXXX

The third email (22) is slightly longer but, upon asking the author of the message whether she was familiar with the notion that words written in capital letters are an indication of shouting, she answered in the negative (which supports the assumption in Section 2.4, Rules and Conventions: Netiquette, that people who use email for private purposes might not necessarily be familiar with the rules and conventions of netiquette).

According to Table 2-3, 15% of the emails are all lower case, whereas 33% make use of correct upper and lower case. The remaining 51% are incorrect in their use of case, i.e. either use lower case instead of upper case (cf. examples (23) and (24)) or, for the sake of emphasis, contain words which are completely spelt with capital letters. Capital letters are a legitimate means of emphasis and not taken amiss as long as there are only a few words capitalised, as examples (25) and (26) illustrate:

14 Cf. netiquette: "you should not TYPE IN ALL CAPITAL LETTERS, AS IT INDICATES YOU ARE SHOUTING" (Jansen 2002: 270).

Table 2-3: Upper and Lower Case Spelling in Emails

Case	number of emails	%
all upper case	3	1%
all lower case	52	15%
correct upper and lower case	114	33%
inconsistent use of capitalisation/minusculisation	173	51%
total	342	100%

Table 2-4: Inconsistent Use of Capitalisation/Minusculisation

Case	number of words	%
lower case instead of upper case	1273	84%
upper case instead of lower case	91	6%
upper case for emphasis (words)	152	10%
total	1516	100%

(23) Bloody hell this *g*entoo looks a pain to install!!!!! *i*m just reading the install file now.. *i*m installing it from the liveCD things… Still waiting for my laptop so *i*m just playing about with it… (My emphasis)

(24) *w*here do u live and stuff? *u* don't wanna run *g*entoo for a server btw.. Slackware or BSD is better. I'll explaine later.. *a*t work atm (My emphasis)

(25) I'm really glad to be home, it is *BOILING* here, I landed at half 10 last night my time and it was some 29C!!!!! (My emphasis)

(26) I like the "*ALL YOURS BABY*!!!", didn't bother me anyway if you had to keep up with this other guy. (My emphasis)

Of those 1516 words which are wrong in terms of case, 84% contain lower case instead of upper case letters, 6% upper case instead of lower case letters, and 10% of the words are entirely capitalised, as Table 2-4 points out. This result clearly shows that there is a tendency to minimise capitalisation, an observation which partly supports Thurlow's (2001: 288) assumption that the language of computer-mediated communication makes "minimal to no use of capitalisation".

2.5.2.3 Non-Conventional Spelling[15]

Lexical reductions[16] abound in email messages. There are clippings, initialisms, consonant clusters as well as letter and number homophones or ad-hoc abbreviations. Furthermore, there are words which are spelt in such a way that they approximate their phonological value, while others resemble comic expressions or emulate prosody. Table 2-5 summarises the examined non-conventional spelling features analysed in the corpus:

Table 2-5: Non-Conventional Spelling in Emails

Spelling	number	per 1000 words
letter/number homophones	264	5.87
g-clippings	15	0.33
consonant spelling	35	0.78
omission of apostrophes	278	6.19
emulated prosody / onomatopoeic exclamatory spelling	560	12.46
phonological approximation and other non-standard shortenings	261	5.81
total	1413	31.44

a) Letter and Number Homophones

Shortenings can be achieved by the use of a single letter whose phonological content is equalled to a word, i.e. by the use of letter homophones. The letter /u/, for instance, refers to the pronoun 'you', the letter /c/ to the verb 'see', the letter /r/ to the finite verb form 'are', and the letter /y/

15 In this study, the term non-conventional or non-standard spelling or shortenings is used as a general label for words or abbreviations whose spelling does not correspond to the orthography of the major dictionaries such as the *Longman Dictionary of Contemporary English*. Informal or spoken words that have already made their way into these dictionaries are excluded. Similarly, acronyms and common initialisms are not analysed, as many of them, such as *asap*, have been generally acknowledged and are also mentioned in the *Longman Dictionary of Contemporary English*.

16 Common abbreviations are excluded from the analysis, as they would not contribute much to the linguistic characteristics of email.

to the question word 'why'. Letter homophones can stand on their own and refer to a single word, but they can also be attached to another letter homophone and refer to two words, as example (27) illustrates where *ru* means 'are you'.

(27) How *ru* getting on? (My emphasis)

Letter homophones can further be combined with number homophones, as quotation (28) shows, in which the letter /b/ is short for the syllable 'be-' and the number *4* for the syllable '-fore', adding up to 'before'.

(28) i had hovered up all downstairs *b4* i left and i know someone had scrubbed the surfaces/cooker too. (My emphasis)

Even though they are characteristic features of computer-mediated communication, neither letter nor number homophones are frequently used in email messages. As Table 2-5 shows, there are only 5.87 instances of homophones per 1000 words. The homophone /u/ is most prevalent: 82% of all the homophones are *u*-homophones, i.e. 216 instances, including words such as *ur*, *urs* and *urself.*

b) *G*-Clippings

In speech, the pronunciation of *ing* in words such as *going*, *loving,* and *darling* constitutes a phonological marker of register and can be pronounced [iŋ] or [ɪn] (cf. Trudgill 1996). Considering examples (29) to (31), this difference in pronunciation may also be evident in emails where the *g*-clipping in *ing*-forms indicates the author's pronunciation of *ing* as [ɪn]. It is further remarkable that the letter /g/ is not substituted by an apostrophe, which, of course, is a matter of economy.

(29) I told [Name] that you'd said about *goin* down and he said it could be quite weird […] (My emphasis)
(30) Im still in the RAF, been what 5 years now and Im *lovin* it! (My emphasis)
(31) *Darlin* […] [j]ust got your message baby. (My emphasis)

There is, however, only a slight tendency to drop the /g/ in *ing*-forms: there are 0.33 instances of *g*-clippings per 1000 words, i.e. 1% of all *ing*-forms are clipped. Yet, even though this feature is rarely observed, it nevertheless contributes to the tendency of non-conventional spelling clusters in email messages and with the further development of the email register, it is not unlikely that this feature will become more marked.

c) Consonant Spelling

Consonant spelling is another effective economic means in computer-mediated communication. "Recognising that consonants usually have more semantic detail/value than vowels" (Thurlow 2002), words are spelt without vowels, as, for instance, the modal verb 'would' in example (32) and the preposition 'from' in example (33).

(32) but thought *wld* reply since u r so strict bout these things! (My emphasis)
(33) lovely to hear *frm* u [...] (My emphasis)

Consonant clusters might as well affect the spelling of /r/ where it is non-syllabic (i.e. post-vocalic) as it is the case in the consonant cluster *gdn* meaning 'garden' (34) or *wds* meaning 'words' (35).

(34) Going to Covent *Gdn*/Carnaby st with [Name] on sat to do some hardcore rare trainer shopping. (My emphasis)
(35) Since essay lengths are pretty short, like 1500 *wds*, by the time you've chucked that in, plus the required quotes, bla bla bla, there's not one original thought in it. (My emphasis)

The reason for the omission of /r/ in consonant clusters is that /r/ cannot be classified as a proper consonant but is "some kind of semivowel" (Ladefoged 1993: 92). Dependent on its position, it can either be syllabic or non-syllabic and tends to be dropped in non-syllabic positions. In syllabic positions, i.e. immediately after a consonant or at the end of a word, the liquid /r/ is treated like a consonant and thus spelt, as examples (36) and (37) prove.

(36) WHEN YOU GETTIN SOME *CRDT*? (My emphasis)
(37) X *yr* fingers at all times till i get in touch to let you know that he's back. (My emphasis)

Again, however, consonant spelling is not very frequent in emails: there are only 0.78 instances of consonant spelling per 1000 words.

d) Apostrophes

Thurlow (2001: 289) points out that computer-mediated communication is generally "blamed for the 'death' of the apostrophe", but specifies this statement in a later article in which he explains that "it does not appear that the apostrophe is dead just yet" (Thurlow 2002). In fact, there is a remarkable tendency to omit apostrophes, for this is a further means of economy. Moreover, letter homophone combinations such as *ur* are never spelt with an apostrophe in between even though their meaning can become rather ambiguous: when the letter homophone /r/ is added to the pronoun *(yo)u*, it immediately turns into a homonym, either meaning 'you are' or representing the second person possessive pronoun 'your'. There are eleven instances of *(yo)ur* meaning 'you are' in the corpus; their actual meaning can only be disambiguated when considering the syntactic context, as examples (38) and (39) show.

(38 a) Hope *your* keeping well and enjoying your hols [...] (My emphasis)
b) so cool that *ur* on hols from 23 oct - 04 nov (even though i unfortunately start on 01. nov again)... (My emphasis)

(39 a) I got *your* text message today. (My emphasis)
b) just a quickie to say thanks for texts, can't wait to hear *ur* details as well, have been out & about, & out again 2nite, but will be in 2moro later on - after about 19.30. (My emphasis)

Apostrophes, however, are not only left out in combination with letter homophones but also elsewhere. Hence, the word *its* instantaneously becomes a homonym if spelt without apostrophe and either refers to the third person possessive pronoun 'its' or the pronoun-verb combination 'it is'. In the analysed corpus, 18% of all the apostrophes are omitted, which corresponds to 6.19 instances of apostrophe deletions per 1000 words.

Surprisingly, there is also a slight counter-tendency observable: it happens occasionally that the plural /s/ is spelt with an apostrophe[17]. There are 14 instances of such plural forms in the corpus – mainly initialisms such as the ones in examples (40) and (41).

17 There seems to be a general uncertainty towards the plural /s/ in acronyms and initialisms so that there even exists an online English language poll on whether one uses an apostrophe in plural acronyms and initialisms or not. It also offers a platform on which people can discuss their opinions. For further interest, see <http://www.usingenglish.com/poll/index.php>.

(40) Hey if we don't jam soon just give *CD's* and funds to [Name]. (My emphasis)

(41) One of them is a south african dude who saved me £10 entry to the club cause he knows alot of the *DJ's* and Organisers so its going to be quite handy knowing him. (My emphasis)

e) Emulated Prosody and Onomatopoeic Exclamatory Spelling

An important aspect of the influence of spoken language on computer-mediated communication is emulated prosody. Similar to today's Short Message Service, which does not (yet) allow the users to give their messages special emphasis by underlining single words, writing them in bold letters or colouring them, it was impossible to use purely graphic emphasis in early emails. Due to the restricted American Standard Code of Information Interchange (ASCII code)[18], which did not contain signs for typographic information, one had to be inventive in the early years of computer-mediated communication to be able to add emphasis to the text. Today, an extended ASCII code, the Rich-Text-Editor, and HTML[19] enable the email users to individualise their messages and layout them in various different ways. The old traditional ways of adding emphasis to words, however, are still maintained and have not been replaced by newer options of text-layouting. This seems to be mainly due to the fact that the extended ASCII code and HTML are platform-dependent and so, transfer errors more easily occur, especially if the receiver's computer software does not support these features.

18 The ASCII code is a table of codes that translates letters in the alphabet (a–z/A–Z), numbers (0–9) as well as special characters such as exclamation marks, question marks, etc. into binary codes. Thereby, each character is always represented by the same order of number(s), e.g. A as 65, B as 66, etc. The standard ASCII code consists of 128 character codes (numbered from 0 to 127), of which the first 32 are non-printable control codes and the remaining 96 are representable characters. In addition to the 128 standard ASCII code, there is another set of 128 codes (known as extended ASCII code). This code, however, is platform-dependent and usually redefined according to the regional configurations to incorporate those symbols which are specific to the language spoken in the area. (Resources Network 2003)

19 HTML email (also known as graphic rich email) is the term for email "that is formatted using HTML, as opposed to plain text, so that graphics and links appear in the email body. In order to see the format and images correctly, both the sender and the recipient must have an email program that supports HTML email" (Jansen 2002: 204).

Extra emphasis, or "emulated prosody" as Beutner (2002) and Haase et al. (1997) term it, can be achieved in various ways: letters of a word can be reduplicated or capitalised as examples (42) and (43) illustrate.

(42) Hey I used your eye gel mask thing and the eye gel on the flight it was *sooooooo* good!!! (My emphasis)
(43) I'm *REALLY VERY SORRY* about this email, I'm acting like a cow. (My emphasis)

Words can also be put in between asterisks[20] or repeated as in examples (44) and (45).

(44) My powers of deduction aren't too good so I'm writing to you both in the hope that i've got the right people! **SIGH** (My emphasis)
(45) It was *really really* nice to hear from you. (My emphasis)

Emulated prosody can further be achieved by underlining the spaces around the word that is to be emphasised, as illustrated in examples (46) and (47).

(46) Do you_*ever*_check your mail then? (My emphasis)
(47) I'm still on £5 per hour and I know I _*need*_ (let alone _*deserve*_) at least 7 or 8. (My emphasis)

A special case of emulated prosody, according to Beutner (2002: 66), is the creative use of punctuation: "multiple question marks or exclamation points signal the reader to intensify the degree of rising intonation in a question or the loudness of an assertion" (Lee 2003: 320). Similarly, trailing dots may indicate a pause as well as silence or even irony. In the analysed email corpus, there are 0.62 instances of repeated question marks per 1000 words, 3.12 instances of trailing exclamation points and 7.92 instances of trailing dots (cf. Table 2-6 below). Whereas question marks are never repeated more than four times in a row (cf. quotation (48)), exclamation marks and full stops can easily

20 It is usually verbstems that are written in between asterisks. These verbstems are used in a predicative way and termed inflectives (Schlobinski 2001: 193) or "Aktionswörter" (Döring 1999: 45), i.e. words that express an act.

be accumulated, as examples (49) and (50) illustrate. There are 0.42 instances of four or more combined exclamation marks per 1000 words and almost twice as many instances of trailing dots (0.78 instances per 1000 words).

(48) hi y do u keep sending me blank emails???? [name]

(49) haven't started my essay yet :-/ but will now!!!!!!!!!!!!!!!!!!!!

(50) You should know why this is.........., got that little situation I told you about to handle as its on my mind all the time too.

Thurlow (2002) differentiates between emulated prosody and onomatopoeic exclamatory spelling. The latter category includes exclamation words such as *yeah!*, *hehe!*, *ahhh!*, etc. that are of an onomatopoeic nature. Thus Thurlow expands the feature of emulated prosody to the word class of interjections. These two categories, however, may overlap, especially where interjections contain iterations of letters or are reduplicated. Yet, in both cases, the accent as prosodic element of the spoken language is expressed in writing (Beutner 2002: 66) and onomatopoeic exclamatory spelling can thus be considered a sub-category of prosodic features. Examples (51) and (52) illustrate the use of onomatopoeic exclamatory spelling in the corpus. Examples (53) and (54) provide two instances of emulated prosody.

(51) Then, we get to go and test the tent out on a proper camp site (finance permitting) *woohoo*!! (My emphasis)

(52) Ooh, I finally got my pictures back. *Yeah!* So things are just peachy here. (My emphasis)

(53) OK? i know i've got to beg at least five times to be allowed to pick u up, so here goes: please, please, please, please, *pleeeeaaaasssssseeeeeee*! Wow, now i'm knackered. (My emphasis)

(54) I hand in my essay on Friday so will be very very willing to go out and *paaaaaaaaaaaaaaaaaaarty*! (My emphasis)

Emulated prosody and onomatopoeic spelling are the most prevalent non-conventional spelling features in the analysed texts, appearing 12.46 times per 1000 words.

f) Phonological Approximation and Other Non-Standard Shortenings
Due to the spoken-like character of computer-mediated communication, there is a tendency to spell the words in such a way that they represent the specific pronunciation of their user. Spellings like *ya* for 'you' (also used in blends, e.g. *cya* (meaning 'see you'), *dya* (meaning 'do you') and *heya* (meaning 'hey you')) or *eva* for 'ever' are attempts to approximate the phonetic value of the specific word. There are 261 instances of such spellings in the corpus, equalling 5.81 instances per 1000 words. Lee (2003: 319) terms this kind of writing "mock oral writing" as the "deviations from formal grammar usually signify effort to visualise talk" but do not actually ensue a change in pronunciation. Hence, "it would make no *phonetic* difference" if the word *eva* in example (55) or the word *kinda* in example (56) were spelt correctly, but "the friendly, casual quality of the text would be lost" (Lee 2003: 319).

(55) Well won't make this too long, cos i could go on for *eva*... (My emphasis)
(56) Hey I was trying to think of what *kinda* music I should send you... (My emphasis)

An already well-established way of spelling the lexeme 'love' is *luv*. Whereas it used to be "an informal spelling of *love* when it is used to address someone" (*Longman Dictionary of Contemporary English* 1995: 858), it is nowadays also written in farewell formulas and elsewhere. Similarly, the word *nite* has become a CMC standard spelling of the lexeme 'night' and as such also appears in compounds or in combination with number homophones like *2nite*. Surprisingly, these two features do not appear very often in the examined email corpus (there are merely six instances of *luv* and four instances of *nite*). Yet, being more characteristic features of text messaging, they will be further discussed in Section 3.5.2.3, Non-Conventional Spelling.

Last but not least, there are some non-standard spellings which cannot be classified easily, as they constitute lexical innovations such as ad-hoc abbreviations or spontaneously created and idiosyncratic ways of shortenings. Quotations (57) to (61) provide typical examples of ad-hoc abbreviations and free variations respectively: the word *boyf* in (57) refers to 'boyfriend', the words *yest* and *tomm* in (58) and (59) mean 'yesterday' and 'tomorrow', while *ofish* in (60) is short for 'officially'.

Another remarkable feature is the spelling of the double letter /e/ by means of the number *3* as in *g33k* (meaning 'geek') and *l33tness* (meaning 'leetness'[21]) in example (61).

(57) [name] i think is coming in a coupld of days, and we are celebrating new year with [name] (who unfortunately it seems might be breaking up with [name] her *boyf...*) (My emphasis)
(58) Spent *yest* trying to find out about companies in Bath to whom i could write and see if they can offer me something. (My emphasis)
(59) i am going down for the wedding *tomm*, and really excited. (My emphasis)
(60) I conclude that ur *ofish* insane. (My emphasis)
(61) I am really a *g33k*, but I needed someone to share the *l33tness* of this with :-D (My emphasis)

2.5.3 Punctuation

Lan (2000: 27) describes the use of punctuation in emails as "carefree" while Thurlow (2001: 288) calls it "creative". There is, in fact, a rather lax attitude towards punctuation marks: they are either employed correctly, omitted or else used in a creative and superfluous way. Sometimes, full stops, commas or exclamation and question marks are replaced by dashes, and whenever emoticons appear at sentence boundaries, punctuation marks – usually full stops – tend to be omitted.

Table 2-6 illustrates the use of punctuation marks in the analysed corpus. Essential punctuation marks are omitted 7.88 times per 1000 words, which can result in asyndetic sentences[22] (cf. example (62)). Sometimes, punctuation marks are also employed inappropriately. This happens 3.63 times per 1000 words. Often, commas are used instead of full stops and other sentence final punctuation marks, which again may result in asyndetic sentences as can be derived from example (63).

---

21 The word *leet* is a slang expression for 'elite'.
22 In contrast to other (more formal) texts which are punctuated correctly, it is often difficult in emails (as well as in text messages) to decide whether asyndetic sentences simply result from omitted or wrongly applied punctuation or whether they are, as this is the case in correctly punctuated texts, the result of lack of a conjunction. Asyndetic sentences will be further discussed in Section 2.5.4.1, Syntactic Reductions.

(62) its like -22 degrees today with the windchill am a frozen popsicle [...]
(63) how are your exams going , u are gonna top all ur papers , touchwood , so chill out girl and have fun while u are at it...

Table 2-6: Punctuation Marks in Emails

Punctuation marks	number	per 1000 words
dashes instead of full stops, commas, question or exclamation marks	59	1.31
inappropriate punctuation marks	163	3.63
omission of essential punctuation marks	354	7.88
question marks and exclamation marks combined	11	0.24
trailing dots (full stops)	356	7.92
trailing exclamation marks	140	3.12
trailing question marks	28	0.62
single exclamation marks	563	12.53
single full stops	1777	39.54
single question marks	386	8.59

As for the creative use of punctuation, full stops as well as exclamation and question marks appear 11.66 times per 1000 words in an accumulated way. Trailing dots occur most frequently, namely 7.92 times per 1000 words, followed by the repetition of exclamation (3.12 times per 1000 words) and question marks (0.62 times per 1000 words). Examples (48) to (50) above provide an example of each. There are 39.54 instances of single full stops per 1000 words, 12.53 instances of single exclamation marks per 1000 words and 8.59 instances of single questions marks per 1000 words. Examining these results, it is noticeable that ordinary statements followed by an exclamation mark are rather common. Compared to the frequency of exclamation marks (15.65 instances per 1000 words), full stops appear only about three times as often (47.46 instances per 1000 words). The printed email in quotation (64) (with sentences numbered) provides an example of such frequent use of exclamation marks: in fact, apart from sentences (1), (2), (8), (9), (13), and (17), every sentence ends in an exclamation mark, and five of the eleven sentences ending in an exclamation mark contain trailing exclamation points, these being sentences (3), (6), (7), (11), and (15).

(64) (1) Hey sweetie,
As you've probably figured I am home safe and sound. (2) My flights were fine, I ate ALL the meals they gave us, yet when I went to bed, I was still hungry.. (3) i'm becoming like you!! (4) (just joking!) (5) Its half 5 in the morning here, and I'm really jetlagged, just can't sleep so might put on a video for a bit or find some food!
(6) Hey I used your eye gel mask thing and the eye gel on the flight it was sooooooo good!!! (7) I looked the coolest out of everyone there on the flight, then again don't I always!!!! (8) So how was friday night? (9) I want details.... (10) how was the meal, the night at [names], hope you took pictures to show me!
(11) I'm really glad to be home, it is BOILING here, I landed at half 10 last night my time and it was some 29C!!!!! (12) I'm not complaining! =) (13) Anyways, best be off and try to get some sleep.... (14) oh before I forget, give me the name of your hair serum that you get from ch, will see if I can get it here and will see if its cheaper,chances are it might well be! (15) Take care honey, i will be in touch, I can't believe you'll be skiing this time next week WOO HOOOOO!!!
(16) Lots and lots of love to you and all in the house! (17) (who's left anyway?)
love xxx, [name]

Another finding concerns the tendency to omit sentence-final punctuation when emoticons are inserted. 107 of the 142 emoticons in the corpus, i.e. 75%, thus substitute such punctuation marks, mainly full stops or, after farewell formulas, also exclamation points. Whenever sentences end with a punctuation mark other than a full stop, emoticons tend to appear in combination with them. Examples (65) to (68) illustrate this finding.

(65) I start work at 7AM this time and work though till 3PM and I communicate with work via ICQ so I'm on almost all the time if you wanna chat :-)
(66) [Name] sent me 2 emails from british people.. I'm guessing his email system isn't awefully well setup :-/
(67) I'm ganna install gentoo.. Just reading the install guide... looks like it ganna take ages! :(
(68) Hi Guys, find attached!? :)

Accordingly, it could be hypothesised that emoticons can also serve a certain punctuation function insofar as they can either replace full stops or, after farewell formulas, exclamation marks. This assumption cannot be proved at this stage, however, as, on the one hand, there are not enough examples available in the email corpus and, on the other hand, this might as well be an idiosyncratic tendency of the analysed corpus. To see whether this hypothesis can be verified and also bears the potential of universal value, the punctuation function of emoticons will be more closely analysed in the SMS corpus as well as in the German and Swiss-German corpora. The results will be discussed in Section 5.3, Emoticons as Punctuation Marks.

2.5.4 Grammatical Features

2.5.4.1 Syntactic Reductions

Syntactic reductions are a very central and characteristic feature of computer-mediated communication. Encouraged by the factor of economy and speed with which messages are composed, sentences tend to lack various syntactic features and come to resemble the so-called telegram style[23], as can be seen in example (69).

(69) HI sweety,
Didn't get any mail from you, so don't know where it went.
Was feeling a bit crap yesterday with exam revision so just wanted to chat to you, was going to ring but then got caught up with work.
Hows you? And how come all the high bills??? Because of me???
Gotta go, have an exam to revise for,
You take care [name], and do mail me, will check asap!
Love you lots,
[Name] xxx

The most common occurrences are the deletion of subject pronouns, conjunctions, finite forms of the verb *to be* (also in combination with

23 For further reading, cf. also Beutner (2002: 70) or Bittner (2003: 172).

the omission of existential *there*), articles or possessive pronouns[24], as well as the omission of auxiliary verbs. There are 35.78 elliptic phrases per 1000 words as Table 2-7 illustrates.

Table 2-7: Elliptic Phrases in Emails

Ellipses	number	per 1000 words
articles / possessive pronouns	123	2.74
auxiliary verbs (without subject deletion)	31	0.69
be as a main verb (copula deletion) (of which existential *there*/article + copula deletion)	127 (46)	2.83 (1.02)
conjunctions	496	11.04
infinitive marker *to*	12	0.27
prepositions	14	0.31
pronouns other than subject pronouns	19	0.42
subject pronouns	572	12.73
subject pronouns and auxiliary verbs	168	3.74
total	1608	35.78

According to Ochs, subject deletions "occur frequently and systematically in unplanned discourse, but in planned discourse they do not occur" (Ochs 1979 quoted from Milroy 1985: 146). Hence it is not surprising that the majority of elliptic phrases in the corpus is marked by subject deletion (12.73 times per 1000 words), mostly first person deletion as this pronoun abounds in interactional emails: "[e]mail letters have a clear-cut-me-perspective. The producer writes his or her email in the first person singular so that predicative verbs often occur in performative mode" (Uhlirovà 1994: 278). Other deletions – such as second person pronoun deletions – also occur. The second person pronoun *you*, for instance, can be left out in questions addressed to

24 Often, it cannot be decided whether a noun lacks its article or a possessive pronoun, so articles and possessive pronouns are not distinguished here.

the receiver as example (70) illustrates (the gaps are marked by ø and øø respectively when two words are omitted). Example (71) lacks all its first person singular pronouns *I*, whereas example (72) proves that subject deletions may occur irregularly and do not follow a specific pattern. Again, a remarkable inconsequence is to be observed: in one text sentences can intermittently alternate between being syntactically complete and incomplete.

(70) øø Living anywhere new? øø Fired anything new? ;)
(71) ø Didn't get any mail from you, so ø don't know where it went. ø Was feeling a bit crap yesterday with exam revision so ø just wanted to chat to you, ø was going to ring but then ø got caught up with work.
(72) thanks for the mails. i would love to ask [name]. ø doubt he will be able to because of exams etc, but ø will let you know.

Yongyan (2000: 32) states that "[t]hese ellipses reinforce the spoken touch in emails". Often, they also appear in a complex form where several parts of a sentence are omitted at once, e.g. the omission of the subject and auxiliary verb as in example (70) above or example (73) below.

(73) will you go to the chinese disco? øø forgotten when it will take place tho'.

Such subject-auxiliary-verb ellipses occur 3.74 times per 1000 words. Auxiliary verb deletion, however, does not necessarily include the omission of the subject: there are a few auxiliary verb deletions that are not accompanied by the ellipsis of the subject (0.69 instances per 1000 words), which is a further feature characteristic of the conceptually oral language of emails. Quotations (74) and (75) provide two examples of such auxiliary verb deletions.

(74) What ø you been up to jobwise these days?
(75) WHEN Ø YOU GETTIN SOME CRDT?

Another feature is the omission of definite or indefinite articles and possessive pronouns. Ferrara et al. (1991: 20) explain that it is well known that "articles are not necessary for human language, and it is therefore not surprising that they disappear in interactive written discourse since definiteness or indefiniteness can be inferred from context". This requires further comments. Often, nouns with a missing (in-)definite article might lack a possessive pronoun instead. In example (76), for instance, it is not clear whether the definite article is omitted or whether the noun *hols* lacks a possessive pronoun:

(76) The first week of ø hols I spent in Wales with [Name], trying to sort things out between us.

It can, however, be argued that possessiveness may be deduced from context: hence, disregarding the possibility of a possessive pronoun, it is obvious that the sender is writing about her holidays, not somebody else's. There are 2.74 instances of article or possessive pronoun deletions per 1000 words.

Copula deletions, i.e. deletions of *be* as a main verb, are about as frequent as article or possessive pronoun deletions: they occur 2.83 times per 1000 words. It is remarkable that about a third of these deletions occur in combination with either the omission of existential *there* or the ellipsis of an article (1.02 instances per 1000 words). This finding supports Ferrara et al.'s (1991: 20) conclusion that copula deletion often co-occurs with the omission of existential *there* or an article or possessive pronoun respectively. Quotations (77) to (79) depict three different examples of copula deletions: quotation (77) simply lacks the finite form *is*, whereas quotation (78) omits the finite form *are* and the possessive pronoun *your*, while quotation (79) contains an instance of existential *there* deletion which is followed by copula deletion.

(77) Just remember who ø afraid of Virginia wolf?
(78) øø exams over?
(79) øø Takeaway pizza everywhere!

Occasionally, the infinitive marker *to* (0.27 instances per 1000 words) as well as prepositions (0.31 instances per 1000 words), and pronouns other than subject pronouns (0.42 instances per 1000 words) are deleted, too, as Table 2-7 shows. These deletions, even though they are rare, further contribute to the tendency to economise the writing of computer-mediated communication.

A further common element of syntactic reduction, in fact the second most frequent feature, concerns the deletion of conjunctions which results in asyndetic sentences. Before more closely examining asyndetic coordinations, however, a few general remarks about conjunctions have to be made. As it is characteristic of unplanned discourse, sentences in emails are of simple sequences and make frequent use of the coordinating conjunction *and*, as Table 2-9 shows. *And* as coordinating conjunction occurs 13.64 times per 1000 words, followed by *but*, which only occurs half as often as *and*, i.e. 6.14 times per 1000 words. This result supports Uhlirovà's (1994: 278) finding that "[e]ven in longer email letters, sentences follow one after another usually without explicit intersentential connectors. The most frequent connector used is the most general and semantically more or less redundant *and*". Uhlirovà (1994: 278) further states that "[m]ore specified intersentential semantics remain implicit, as a rule. This feature makes email letters nearer to speech than to written texts, for which overt expressions of intersentential relationships are much more typical". Accordingly, subordinating conjunctions are less dominant as they are characteristic of informational discourse (Biber 1988: 107). Yet, there are nevertheless two types of subordinators that are typical of unplanned and involved discourse, namely causative[25] and conditional[26] subordination. They "seem to be associated with a relatively loose presentation of information due to real-time production constraints, and they seem to mark a range of affective functions relating to the elaboration of personal attitudes or feelings" (Biber 1988: 107). This finding "runs counter to the expectations of many previous studies, which have claimed that all dependent clauses are syntactically complex and therefore reflect informational elaboration [...] [yet] certain kinds of subordination represent a relatively fragmented rather than integrated packaging of information, and [...] subordination can be characteristic of non-informational (spoken) discourse" (Biber 1995: 144).

Table 2-9 shows that, compared to other subordinating conjunctions, causative as well as conditional subordinating conjunctions appear frequently: 59% of all the subordinating sentences are made up of causative or conditional conjunctions. They can be considered "markers of affect

25 Causative subordinators include the conjunctions *as* and *because*.

26 Conditional subordinators include the conjunctions *if* and *unless*.

Table 2-8[27]: Conjunctions in Emails

Conjunctions	number	per 1000 words
asyndetic	496	11.04
coordinating	1091	24.28
subordinating	651	14.49

Table 2-9: The Most Frequent Conjunctions in Emails

Conjunctions	number	per 1000 words
coordinating conjunction *and*	613	13.64
coordinating conjunction *but*	276	6.14
other coordinating conjunctions	202	4.50
causative subordination	217	4.83
conditional subordination	168	3.73
other subordinating conjunctions	266	5.92

or stance, that is, as justification for actions or beliefs ([*as*], *because*) or conditions for actions or beliefs (*if*, *unless*)" (Biber 1988: 107).

The lack of coordinating or subordinating conjunctions results in asyndetic sentences[28]. Asyndetic sentences, such as the ones in examples (80) to (82), are a recognised feature in the analysed corpus: they occur

27 The data excludes subordination features such as *wh*-clauses, sentence relatives and *that*-complementisers. These are already well examined characteristics of unplanned and involved discourse, and as the focal point is asyndetic sentences, they do not require any further elaboration. *That*-complementisers are not included in this analysis for they cannot actually be assigned to subordinating conjunctions. The *Longman Dictionary of Contemporary English* regards *that*-complementisers as facultative (e.g. "Harry thought [that] it was a lousy idea.") and so, even though Biber (1988: 244) classifies them as subordinators and considers their deletion a form of syntactic reduction and thus typical of speech, their deletion cannot be regarded as a form of syntactic reduction that is characteristic of orality only.

28 Due to the incorrect use of punctuation in computer-mediated communication, asyndetic sentences can also result from omitted punctuation marks such as full stops (cf. Section 2.5.3, Punctuation).

11.04 times per 1000 words and, according to Table 2-8, are almost as frequent as subordinating sentences which occur 14.49 times per 1000 words.

(80) Hi! Where's richmond? Tomorrow I am in class until seven, would love to party after, giving in essay tomorrow! Where and when?! Xx
(81) That's right you should reply, still, hope spanish was fun, hola!
(82) Everyone here is fine, I have already commenced on the shopping, boughts Ralph Lauren's Romance which smells really nice, also bought some other random things.

As already pointed out above, asyndetic sentences are a feature of economy and as such resemble brief notes: what comes to mind is quickly written down without being structured. Thereby, linking words are neglected, and so are other stylistic devices. Yet, the messages remain understandable, since readers can easily fill in the missing semantic information, which is usually a coordinating conjunction such as *and* or *but*, or else some form of causative or conditional subordination.

2.5.4.2 Verbs

According to Biber (1988: 105), private verbs (also called mental verbs) belong to the most characteristic features of Dimension 1. They can be considered as interactive or involved, for they "are used for the overt expression of private attitudes, thoughts and emotions". Table 2-10 shows that mental verbs, in fact, belong to the three most frequently used types of verbs in the analysed email corpus.

Mental verbs make out 15% of all the verbs and appear 22.52 times per 1000 words. They are almost as frequent as activity verbs, which appear 23.88 times per 1000 words, i.e. 16% of all the verbs are activity verbs. Activity verbs show a slightly higher frequency than mental verbs, which backs Biber et al.'s (1999: 365) observation that "overall, activity verbs occur much more commonly than verbs from any other semantic domain". Both mental and activity verbs are the two most frequent types of verbs in conversation and again, they most often appear in the email corpus besides *be* as a main verb. This finding fundamentally contributes to the conversation-like nature of this register.

Table 2-10[29]: Verbs in Emails

Verbs	number	% of verbs	per 1000 words
activity verbs	1073	16%	23.88
be as a main verb	1595	23%	35.49
communication verbs	247	4%	5.50
mental verbs	1012	15%	22.52
obligation/necessity modals	164	2%	3.65
permission/possibility/ability modals	371	5%	8.26
volition/prediction modals	516	8%	11.48
pro-verbs (e.g. *do, get, put*)	414	6%	9.21
others	1406	21%	31.29
total	6798	100%	151.27

Pro-verbs also commonly appear (9.21 times per 1000 words) and are a similarly important element of Dimension 1. They are characteristic of the involved discourse and as such typically used in speech: "in conversation, where there is pressure to find your words speedily, pro-verbs tend to occur frequently, in part because they save the time that would be needed to find a more explicit verb" (Finegan 1999: 349). The same applies to email messages: since authors of emails write their messages in the real-time mode, pro-verbs such as *do*, *get* and *put*, which substitute "a fuller verb phrase or clause" (Biber 1988: 106), are bound to be used.

Equally common are volition/prediction modals (11.48 instances per 1000 words) as well as permission/possibility/ability modals (8.26 instances per 1000 words). The most typical modals of conversation, according to Biber et al. (1999: 489), are *can/could* as well as *will/would.* Furthermore, *be going to* is the most often used semi-modal in conversation. Among the less frequent obligation/necessity modals, the semi-modal *have to* is named the most common one in conversation. These

29 Auxiliaries (e.g. finite forms of the verbs *to have* and *to be*) are not included in the statistics.

facts are mirrored in the analysed corpus where *can/could*, *will/would*, *be going to* and *have to* achieve the highest scores among modals, as Table 2-11 illustrates.

Table 2-11: Modal Verbs in Emails

Modals	number	% of modals	per 1000 words
permission/possibility/ability			
can	220	21%	4.90
could	77	7%	1.71
may	46	4%	1.02
might	28	3%	0.62
obligation/necessity			
be supposed to	2	0%	0.04
have to	73	7%	1.62
must	12	1%	0.27
need to	20	2%	0.45
should	57	5%	1.27
volition/prediction			
be going to	118	11%	2.63
shall	10	1%	0.22
will	267	25%	5.94
would	121	12%	2.69
total	1051	100%	23.39

Yet, by far the most frequent verb is *be* as a main verb; it occurs 35.49 times per 1000 words. It is "typically used to modify a noun with a predicative expression, instead of integrating the information into the noun phrase itself", and it is "associated with a fragmented presentation of information, resulting in low informational density" (Biber 1988: 106). This also applies to emails: since emails are composed spontaneously or under real-time conditions, they tend to be list-like, containing lists of ideas and main clauses respectively, which are not directly linked with each other and thus show little textual coherence.

2.5.5 Lexical Features

2.5.5.1 Prefatory Expressions

In a conversation, an utterance commonly "opens with a single prefatory word, which has the function of orienting the listener to the following utterance, especially in relation to what has preceded" (Biber et al. 1999: 1074). Such prefatory words can include inserts, i.e. discourse markers, interjections or response forms, as well as adverbs such as stance or linking adverbs (Biber et al. 1999: 1074). Prefatory expressions are a feature of the spoken language but also figure in email messages, which clearly indicates the close relation of the register of email to speech. Discourse markers[30], as Biber (1988: 241) states, "are very generalised in their functions and rare outside of the conversational genres". They "are generalised markers of informational relations in a text" and "help to maintain textual coherence when a text is fragmented and would otherwise be relatively incoherent" (Biber 1998: 106). Email messages can, in fact, be rather fragmented as has been pointed out above (cf. Section 2.5.4, Grammatical Features) and discourse particles clearly serve as a means of structuring the text. Table 2-12 shows that discourse markers appear 1.05 times per 1000 words.

Corresponding to Biber et al.'s (1999: 1083) theory, the term interjection is here applied to inserts with an exclamatory function that express the speaker's emotion. Interjections appear 5.47 times per 1000 words; the interjection *oh* (also in combination with other inserts such as *oh no* or *oh well*) occurs most frequently: 21% of all the interjections are made up of the insert *oh*. The most often used prefatory expression, however, is the linking adverb *so*, occurring 6.21 times per 1000 words. Less common are stance adverbs (1.96 instances per 1000 words) as well as response forms (0.80 instances per 1000 words). Response forms mainly appear in email dialogues, but as the analysed corpus does not contain

30 There is no general list of discourse markers and those elements included could certainly be complemented by other possible discourse markers (cf. Jucker and Ziv 1998). For this analysis, Biber et al.'s (1999: 1086) list of discourse markers is taken over, namely *well*, *right*, and *now* as well as the finite formulas *I mean*, *you know*, *you see*, *mind you*, and *now then*.

Table 2-12: Prefatory Expressions in Emails

Prefatory expressions	number	per 1000 words
discourse markers	47	1.05
interjections	246	5.47
linking adverb *so*	279	6.21
response forms	36	0.80
stance adverbs	88	1.96
total	696	15.49

dialogues of more than two threads, response forms are rare. All in all, however, it can be concluded that prefatory expressions are an important feature of email messages: they occur 15.49 times per 1000 words and significantly contribute to the speech-like quality of the register.

2.5.5.2 Spoken Vocabulary

Spoken words, as the term implies, and regionalisms as well as slang expressions are typical elements of speech, elements which also play a role in email messages. They greatly contribute to the informal and relaxed tone of email and bring it closer to conversation. Finegan (1999: 338) states that "slang has a legitimate place in the linguistic repertoire of speech communities". The use of such informal vocabulary clearly indicates that users do not make the effort to carefully choose their words; being under real-time constraints, they do not have the time to think about the most appropriate lexical items when composing an email, but simply key in what comes to mind first.

Similar to non-conventional spelling, which contributes to the friendly and casual tone, the use of informal vocabulary makes email messages more personal. Since regionalisms and slang expressions are used individually, they tell the reader something about the writer's surroundings, e.g. where he comes from and what social background he has, and thus give the message a personal touch. Quotation (83) provides an example of regionalisms, namely *wee* meaning 'very small' in Scottish English, while quotation (84) illustrates the use of slang expressions.

(83) And yet, I ramble, and I only intended a *wee* hi, so a shall do a *wee* bye instead. (My emphasis)

(84) I'm doing ok, I spent pretty much all of thursday catching up on sleep, plus was feeling a bit *crappy*, missed bombay and [name] a bit too much!!!! (My emphasis)

Other words or structures that are common in conversation are, for instance, the dialectal degree modifier *pretty* in example (85) (cf. also example (84) above) or the sentence-final linking adverbial *though* in example (86) (cf. Biber et al. 1999: 850).

(85) The waether has turned today and it is windy and *pretty* chilly. (My emphasis)

(86) I must remember to make him wash his hair before you meet him *though*... (My emphasis)

Table 2-13 illustrates how frequently each of these elements occur. Spoken words and structures are by far the most common elements, appearing 5.94 times per 1000 words, followed by slang expressions (4.34 instances per 1000 words) and finally taboo expletives (0.45 instances per 1000 words) and regionalisms (0.29 instances per 1000 words).

Table 2-13: Spoken Vocabulary in Emails

Spoken vocabulary	number	per 1000 words
regionalisms	13	0.29
slang expressions	195	4.34
spoken words/structures (excluding prefatory expressions)	267	5.94
taboo expletives (excluding moderated expletives)	20	0.45
total	495	11.01

Taboo expletives are not used very often as Table 2-13 indicates, but they are still worth noticing. Since emails are of a dynamic character and can be written, deleted and re-written quickly, people write more spontaneously. This, however, leads to an ambivalent result: on the one hand, this spontaneity tends to make people more honest, but on the other hand, encouraged by the anonymity of the medium, it also lowers their respect towards each other (cf. Döring 1999). Janney (1996: 202) argues that email is a quick medium in which "the partners are 'close': easy to get in touch with, easy to exchange information with, easy to be friendly with, and if problems arise, easy to be unfriendly with". Some researchers have indeed found that

> the lack of visual audience and paralinguistic cues accompanying the casual speech-like setting lead to "flaming". These sociologists have shown that participants freed from the sociolinguistic conventions of more familiar speech and writing events use language that is more emotionally charged and direct than in other settings (Murray 1990: 44).

The emotional impact and casual writing style of email messages has already become evident in the use of prosodic features as well as in the various kinds of non-standard spellings or in the use of spoken vocabulary. Writing that is emotionally loaded, however, not only expresses happiness and satisfaction, but also anger or annoyance. Swear words as well as expletives constitute a suitable way to work off frustration. Hence, offensive language is not an unknown feature in email messages even though it is infrequent in the corpus analysed in this study.

Murray (1990: 45) comes to a similar result and explains the rare instances of flaming in her corpus with the fact that "the sociological studies were primarily of students and faculty [...] whereas [her] study has been primarily focussed on IBM, a corporation with a different cultural milieu". Yet, similar to the sociological studies described above, my research took place among students and this might therefore be an inappropriate explanation. In view of that, Kiesler's claim that computer-mediated communication such as emailing has been associated with flaming, insensitivity and vulgar behaviour (cf. Kiesler 1987 in McCormick and McCormick 1992: 381) cannot be supported and has to be denied from the point of view of this study as well as from Ferrara et al.'s (1991: 26) point of view who found "no inflammatory language" in their corpora.

Whenever expletives are used, their use can mainly be attributed to the fact that email messages may be emotionally loaded: people write while being in a temper and submit their messages unedited, not necessarily realising what offensive language they have actually used.

Whether taboo expletives or other lexical features characteristic of the spoken language, all of them contribute to the informality of the register of email and computer-mediated communication in general. In this context, Crystal (2001: 127) explains that

> there is an understandable tendency to think of emailing solely in terms of informality. It feels temporary, indeed, and this promotes a sense of the carefree. Messages can be easily deleted, which suggests that their content is basically unimportant. Because of its spontaneity, speed, privacy, and leisure value, email offers the option of greater levels of informality than are found elsewhere in traditional writing.

2.5.6 Emoticons as Paralinguistic Modifiers

Computer-mediated communication can be considered as incomplete for its lack of paralinguistic cues. Even though writing fails to provide these cues in general, hand-written letters provide the reader with still more information about the author and his circumstances. The handwriting, the chosen paper, the colour of the pen, the envelope or the smell of the paper tell the reader much about the situation in which a letter was written. All these cues are non-existent in emails[31], but are partly made up by features such as emulated prosody (cf. Section 2.5.2.3, Non-Conventional Spelling) or glyphs with which certain phrases are annotated and which are either called smileys or, in more general terms, emoticons (Beutner 2002: 73). The term emoticon is a blend of 'emotion' and 'icon'. Emoticons are sideways smiley faces that are composed of typographic symbols such as commas, full stops, and parentheses. They are used in form of tokens (Beutner 2002: 73) and indicate humour, irony, anger, happiness, and sadness as well as various other emotions.

Emoticons have been used for around two decades and were invented by the American Scott E. Fahlman, a principal research computer scientist at Carnegie Mellon University in Pittsburgh. It was in the 1980s

31 Meanwhile, newer email programs like Incredimail make it possible to individualise one's message by choosing various different kinds of email backgrounds, inserting animated emoticons, 3D and sound effects, handwritten signatures, personal voice messages or various graphics. Such email programs, however, are not widely used and thus not taken into consideration in this study.

that "the Computer Science Community at Carnegie Mellon was making heavy use of online bulletin boards [...] these were a precursor of today's newsgroups, and they were an important social mechanism [...] [to] discuss the weighty matters of the day on an equal footing" (Fahlman 2002). Yet, the problem of these bulletin boards, or bboards as they are commonly called, was that, due to the lack of paralinguistic cues, irony and jokes were not always obvious at a first glance and often caused misunderstandings and differences. This problem was taken seriously by the users of the bboards and caused some of them to suggest that "it would be a good idea to explicitly mark posts that were not to be taken seriously" (Fahlman 2002) whereupon "various 'joke markers' were suggested and in the midst of discussion", Fahlman (2002) adds: "it occurred to me that the character sequence :-) would be an elegant solution [...] [and] I also suggested the use of :-(to indicate that the message was meant to be taken seriously, though that symbol quickly evolved into a marker for displeasure". This, in fact, was the birth of the emoticons, dated 19 September 1982. The original message[32] posted by Fahlman on this day reads as follows:

> 19-Sep-82 11:44 Scott E Fahlman :-)
> From: Scott E Fahlman <Fahlman at Cmu-20c>
>
> I propose that the following character sequence for joke markers:
>
> :-)
>
> Read it sideways. Actually, it is probably more economical to mark things that are NOT jokes, given current trends. For this, use
>
> :-(

"This convention", Fahlman (2002) writes, "caught on quickly around Carnegie Mellon, and soon spread to other universities and research labs [...] [and] within a few months, we started seeing the lists with dozens of 'smilies': open-mouthed surprise, person wearing glasses, Abraham Lincoln, Santa Claus, the pope and so on". Today, these

32 The entire thread of the messages, i.e. the "Original Bboard Thread in which :-) was proposed", can be found here: <http://www-cgi.cs.cmu.edu/~sef/Orig-Smiley.htm>.

Table 2-14: Emoticons and their Meaning(s)

Emoticons	translation
:-*	– user has just eaten something sour – smiley after eating something bitter – kissing smiley
:~i	– smoking smiley
:-o	– oops! / uh oh! – smiley orator – talkative smiley – surprise – smiley singing national anthem – shouting smiley
:-p	– smiley sticking its tongue out
:-q	– smiley trying to touch its tongue to its nose – user is a smoker – talking or shouting with cigarette or cigar
:-r	– smiley sticking its tongue out
:-x	– user's lips are sealed – a big wet kiss – user is wearing a bow tie

smileys[33] are still in use and the lists have steadily grown longer. In fact, the Internet is full of lists and smiley dictionaries, but most of the smileys itemised are not generally known: they might not be understood by the receiver and are thus not used. The problem is not only that they are grapho-stylistically too complex to be decoded, but smileys may

33 It must be noted at this point that the sideways emoticons are not universal: users in Japan have their own unique set of smileys and create them "according to their culture, faces and body language" (Jansen 2002: 472). In contrast to the Western smileys, the Japanese ones are "intricate in their design, somewhat ambiguous in their expression and [...] right side up instead of sideways. The basic smiley in Japan, (--), is much easier to recognise as a face than the Western version. But since the mouth does not curl upward (there is no character on the keyboard that can do that), the Japanese smiley is somewhat harder to understand without knowing the context" (Pollack 1996). ^o^ signifies a happy smiley, while an excited one looks as follows: *^o^*.

as well be interpreted differently or else, the same emotional content can be expressed by more than just one smiley. In other words, smileys are not isomorphic as they do not have a one-form, one-meaning correspondence as Table 2-14 clearly illustrates. And how could one know that C=}>;*{)) signifies a drunk, devilish chef with a toupee in an updraft and a moustache and double chin? Hence, the most common smileys still are the original :-) and :-(as well as the winky variant ;-). Accordingly, the basic smiley :-) is the most frequent type in the analysed corpus, as Table 2-15 indicates.

Table 2-15: Emoticons in Emails

Emoticons	translation	number	per 1000 words
=)	happy smiley	2	0.04
*8-)	happy smiley with long hair and geek glasses	4	0.09
:(	frowning/sad smiley	3	0.07
:-(	frowning/sad smiley	11	0.24
:)	happy smiley	16	0.36
:-)	basic smiley for joking statements / happy smiley	65	1.45
:-/	sceptical smiley	8	0.18
:-D	laughing smiley	4	0.09
:-P	smiley sticking its tongue out	1	0.02
:-p	smiley sticking its tongue out	2	0.04
;)	winky smiley for flirtatious/sarcastic remarks	7	0.16
;-)	winky smiley for flirtatious/sarcastic remarks	17	0.38
8-(	sad smiley (wearing (sun)glasses)	1	0.02
8-((	very sad / frustrated smiley (wearing (sun)glasses)	1	0.02
total		142	3.16

Even though smileys have come to be known to be a typical feature of computer-mediated communication, they do not appear very often in the analysed corpus: there are only 142 instances of emoticons, which is 3.16 occurrences per 1000 words. This finding supports Petrie's (1999: 28) observation that, contrary to the widely assumed popularity of emoticons, smileys are not a distinctive feature of email as they are not very frequently used in this medium. The reason for the relatively rare use is difficult to elicit and to a certain extent user-idiosyncratic. Beutner (2002: 78) explains that newbies, i.e. people that are new to the Internet or computers, use emoticons more often than experienced users. This hypothesis, however, cannot be supported from the point of view of this research, as not only experienced users, but also computer geeks inserted emoticons. Hence, in order to find out about the motivations of the users to annotate their phrases with smileys, further research would be required. Any possible reasons or causes for the surprisingly few occurrences of emoticons in this and other email corpora remain a matter of speculation at this point.

3 Text Messages

3.1 Definition

The Short Message Service "is a globally accepted wireless service that enables the transmission of alphanumeric messages between mobile subscribers and external systems such as electronic mail, paging and voice mail systems" (ADC Telecommunications 1999). As such it can be considered as

> an advanced form of paging, but has a number of advantages. If the phone is switched off, or out of the area covered by GSM[1], the message is stored and offered to the subscriber when he reappears. This gives much greater confidence that it has been received. Also, the user needs only one piece of equipment (the mobile phone) and the caller needs to know only one number (the mobile phone number) for telephony and paging (GSM MOU Permanent Secretariat 1990 quoted from Taylor and Vincent 2005: 80).

"SMS may be broadly defined as asynchronous, text-based, technologically mediated discourse" (Thurlow 2003). Usually, each short message is restricted to 160 characters, but newer mobile phones enable the user to link up to twelve text messages so that the maximum message length can be increased to about 1800 characters. Any text-based content can be delivered via SMS. Many providers offer extra services such as real-time information about the news, the weather, the traffic as well as

1 GSM, short for Global Standard for Mobile Communication, is "the dominant set of standards used in Europe and parts of Asia for wireless communications" (Jansen 2002: 192). It provided the necessary technological infrastructure to support the general availability of text messaging (cf. Taylor and Vincent 2005: 77). In fact, it was "only the spread of GSM technology [which] enabled the birth of the text messaging culture, which could not have emerged without the technological innovation of the SMS that was originally designed for an entirely different purpose" (Kasesniemi 2003 quoted from Taylor and Vincent 2005: 77).

various other entertainments such as horoscopes or jokes. Apart from plain texts, text messages may also include binary pictures, i.e. pictures which only have two shades of brightness, namely black and white (cf. *Science and Technology Dictionary*), while newer mobile phones even offer the option to paste animations or melodies. Yet, these services are hardly used for various reasons: either users do not know about them or else, if they are aware of such services, may not use them as it is uncertain whether the recipient's mobile phone is enabled to display these features, especially if it is of another make. Lastly, cost is also a factor, for such services are not free and make text messages more expensive. Hence, there is no such SMS in my corpus; the SMS that I have collected at most include ASCII[2] figures and are of plain text.

A relatively recent feature of SMS are delivery reports, an extra service of which the sender can make use if he wishes. Delivery reports tell the sender whether his message was actually delivered or not. If the recipient's mobile phone is switched off, for instance, the message cannot be delivered and, depending on the model and the make of the mobile phone, the sender may receive a message saying "message pending" which changes to "message delivered" once the message has been opened. Dürscheid (2002a: 9) observes that people often call each other to check whether the intended recipient of an SMS has his mobile switched on. As soon as it rings, they hang up and send their message. Delivery reports, however, make such phone calls superfluous, as the sender knows whether his message could be delivered or not.

2 For more details on the ASCII code, see Chapter 2, footnote 18.

3.2 Historical Background: SMS as Side-Product of Other Mobile Communication Services

The dream of wireless communication among people has long been around and is a fundamental desire of the human being. In 1901, the concept of wireless communication became more widespread with what is known as the "Ayrton prediction". British physicist and Professor William Edward Ayrton (1901) predicted a time

> still far away [but] gradually coming within thinkable distance of the realization of a prophecy he had ventured to make four years before [...] when if a person wanted to call a friend he knew not where, he would call in a loud, electromagnetic voice, heard by him who had the electromagnetic ear, silent to him who had it not. "Where are you?" he would say. A small reply would come, "I am at the bottom of a coal mine, or crossing the Andes, or in the middle of the Pacific" [...] Let them think of what that meant, of the calling which went on every day from room to room of a house, and then think of that calling extended from pole to pole [...] a real communication from a distance based on true physical laws.

It was not long until the dream of wireless communication came true: technology developed so rapidly that in April 1973, Martin Cooper succeeded in making his first call on a mobile phone. He was the first person ever to do so. Cooper, the father of the mobile phone, is "credited for creating the cell phone in 1973 while working for Motorola Inc." (Christian 2001), but it was to take almost another two decades until mobile phones could be used to send text messages. In Europe, the first short message is "believed to have been sent in December 1992 from a Personal Computer to a mobile phone on the Vodafone GSM network in the UK" (Buckingham 2001). Even though SMS was actually a side-product of other mobile communication services – as an originally unidirectional system[3], it was used by mobile phone services to send their customers messages (Androutsopoulos and Schmidt 2001: 2) – it

3 Cf. also Mäkelä et al. (2000: 549): "[...] when mobile phones became popular in Finland developers and designers were asking whether Short Message Service (SMS) is needed at all, and who would use it. They expected that if it was needed it would be suitable mainly for professional use."

has now established itself as a major medium of everyday communication in Europe. Still, it "seems striking and somewhat peculiar [...] that messages were not considered to be something people would compose themselves on their mobile phones" (Taylor and Vincent 2005: 80). This also explains why "GSM phones have always been equipped to receive the text messages but even in the late 1990s some mobiles did not have the capabilities to originate them" (Vincent 2004: 4).

When SMS became available in the 1990s, it was generally regarded as a gimmick by most mobile phone users and it was only on the threshold of the new millennium that SMS gained acceptance (cf. Pritchard 2001). What is more, "initially, messages could only be sent between compatible mobile phones on the same network. It was not until 1998 that the incumbent UK operators agreed 'interconnect' arrangements to allow for messages to be sent regardless of the network operator to which the phone was attached" (Taylor and Vincent 2005: 81). The exchange of SMS has dramatically increased ever since: whereas the monthly traffic of short messages in the UK was only around 0.1 billion SMS in 1999, the Mobile Data Association (MDA) announced on 19 December 2003 that the total number of chargeable person-to-person text messages sent across the four UK GSM networks reached 1.77 billion messages during November 2003, which is 59 million per day, compared to 50 million in November 2002 and 40 million in November 2001. Thus it can be said that there has never been a communication technology before which has asserted itself as quickly as SMS (Schlobinski et al. 2001: 5).

According to a MORI and Lycos study that was conducted in Great Britain in 2000, it is mainly young people who use SMS on a regular basis. Their conclusion is that "[w]hen you consider 50% of 15 to 24 years olds using text messaging are doing so daily it certainly appears to be a preferred form of communication". Another study that was carried out in February 2002 by A.T. Kearney similarly shows that sending text messages has become a "daily ritual, especially for many young mobile-phone users" (*PC Magazine* 2002). Thereby, SMS are used for various purposes such as arranging meetings, asking for information, telling the latest news or simply saying hello, while "surveys show that men find it especially tempting to express their feelings by text messaging, especially those tricky ones like 'I love you' and 'I'm sorry!'" (Mander 2000: 7).

It is also the young people who have quickly adapted to the allegedly cumbersome typing of text messages, a major disadvantage of text messaging over emailing as it seems: typing messages on the keypad of a mobile phone is cumbersome and consumes much more time than typing messages on a computer keyboard (cf. also Taylor and Vincent 2005: 78). But for all that, Grinter and Eldridge (2001: 230) come to the conclusion that "teenagers frequently (77 times; 33%) [...] found text messaging quicker than using other media" for which two reasons are named: "first, they have grown accustomed to the interface and have adapted it to their needs; and second, it avoided long conversations". To ease the typing process, most mobile phones have the so-called T9 (text on nine keys) text input function, a predictive text technology developed by Tegic Communications which makes typing less laborious and more efficient. Users only need to press one key per letter, choosing the key with the specific letter and the software then completes the word automatically. In order to type the word 'hello', for instance, the keys on the keypad just need to be pressed five times compared to the 13 times without T9. If a given word is not provided for by the software, there is an option to add it to the predictive dictionary.

The T9 technology does, in fact, simplify the typing process to a great extent, but it has never really caught on for several reasons. Problems arise when the same key-combination corresponds to various different words: the words 'come', 'bone', 'cone' include exactly the same four key strokes, namely 2663, and so do the words 'home', 'gone', 'good' and 'hood', which can be written by keying in 4663. In such cases, one can go on scrolling through possible matches until one finds the word one is looking for or else, type it in without the help of T9 which, however, is arduous and disrupts the typing process. Another drawback of T9 is that once a word is added to the dictionary, it can hardly be deleted, which means that all typos remain. Consequently, T9 is not very commonly used: Grinter and Eldridge (2001: 230) explain that in their study, "[m]ost of the teenagers [...] knew the interface so well that they did not even look at the screen as they typed their messages. Predictive typing, by completing words and often making it difficult to rectify the changes, got in the way of typing rather than supporting it". And *The Economist* (2002) acknowledges that it "is all very clever, but it is such a fiddle that surveys suggest that only about half of people with T9 on their phones actually use it". Brown's findings support these observations: according to her survey,

> only about half (55%) of [the] participants reported us[e] this facility – mainly because it was thought to be quicker and easier. On the other hand, reasons given by those 37% who said they did not use predictive text included, in order of priority, that it was too difficult to use, they did not actually have the facility to start with, it was annoying, it did not choose the right words, it was slower to use and did not facilitate the need for abbreviations (Brown 2002 in Thurlow 2002).

This negative result cannot only be related to the problem described above, but is also due to the enormous spelling variety. The language of text messages is highly individual: words are spontaneously abbreviated and modified, punctuation marks are playfully applied and often, dialect is used. The T9 function certainly enables users to add their own words, but it does not support the spontaneous creation of new words and with it the ludic mode of the medium. Hence, when using T9, users can no longer create playful text messages and give them an individual touch, but must to a great extent correspond with the standard language of the predictive dictionary, its traditional orthography and grammar rules, and thus may experience this function as cumbersome rather than helpful. In other words, T9 is not suitable for any language other than the standard language.

Because of the problems with T9, a superior predictive text technology, called LetterWise, has been developed by Eatoni Ergonomics in New York. Its technology is fundamentally different from T9, as

> it is not based on a dictionary, but on a series of tiny rules that enable it to predict, given what has already been typed, which letter the user is most likely to want next. For example "Q" is most followed by "U" [and] of the three letters "MNO", words most commonly start with "O" [while] the letters "ST" are most often followed by a vowel and so on (*The Economist* 2002).

LetterWise is a very promising technology that also enables users to key in names, addresses, URLs[4] and abbreviations. It is a technology

4 An URL (short for Uniform Resource Locator) is an "acronym/term that describes the location and access method of a resource on the Internet; for example, the URL

that works well but, despite all its advantages, "has failed to take off [because] mobile-phone makers are reluctant to break ranks and defect from T9" (*The Economist* 2002). However, once LetterWise is accepted by the mobile-phone industry, Eatoni can offer an even more ingenious technology called WordWise, which is the most economic program on the market: whereas multi-tap requires 2.2 keystrokes per letter typed, LetterWise reduces it to 1.18 and WordWise to 1.00005 strokes so that, once these technologies are accepted, Eatoni will have reached a major step towards its target, namely to make typing nearly as easy as talking (Mooney 2001).

3.3 Structure

The structure of text messages is of a similarly fixed pattern as that of email messages even though it is much simpler as it consists of a body and signature only. The signature has a similar function as the header in emails: it tells the receiver the mobile phone number of the sender as well as the date and time of submission, and if the message is sent via the Internet, it further indicates the name of the Internet service. There is no option of adding a subject line to a text message, which would be futile anyway as text messages are very short and often do not even make use of all the 160 characters available (cf. Section 1.3, Methodology: Data Collection and Corpora).

Text messages resemble letters even less than emails; still, the sign which signals a new message on the mobile phone is an envelope. This envelope remains closed as long as the message is unread and changes to an open envelope once the message has been opened. The affinity of text messages to letters is remarkable, but a comparison with telegrams and postcards would be more suitable, as both are equally restricted in length and thus exhibit a similar linguistic economy. Schwitalla shows

---

<http://www.netlingo.com> describes the type of access method being used (http – the protocol) and the server location that hosts the Web site (www.netlingo.com – the address)" (Jansen 2002: 390).

that telegraphese[5] is marked by prosiopesis of a sentence's initial material such as articles and subject pronouns as well as by deletion of auxiliaries, copular verbs and prepositions (Schwitalla 2002 in Bergs 2003: 9) and Bergs (2003: 9) concludes that "these are exactly the elements that are also omitted in short message communication". Postcards, like text messages, "exhibit […] those features that are associated with proximity-related language use, such as contractions, short forms, non-standard spelling, non-standard linguistic structures and vocabulary (dialectal, sociolectal)" (Bergs 2003: 10). Both telegrams and postcards, of course, can serve other purposes than text messages, and whereas SMS can be close to synchronous discourse, neither of them is composed under real-time constraints. It is therefore most apt to conclude that SMS is a means of communication in its own right: showing features of various different types of discourse, it is a chameleon-like register which is difficult to categorise and while approximating synchronic communication, it resembles speech more than any asynchronous written communication.

3.4 Rules and Conventions: Textiquette

As the Short Message Service is an independent service and not directly linked to the Internet, netiquette has long had no significance for text messages. In other words, it applied to electronic communications such as email, mailing lists and Usenet, but not to SMS. This also explains why some text messages were (and still are) written exclusively in upper case letters: upper case letters used not to be classified as shouting in this medium and so their use was legitimate. In other words, at the very beginning of text messaging, there seemed to be no generally valid rules or conventions. An etiquette for SMS, the so-called textiquette, has only recently been developed, and so, while the term netiquette has become familiar to most people, the term textiquette has not, nor is it included

5 For a closer comparison of text messages and telegrams, cf. Chapter 7, Text Messages as the Renaissance of Telegrams.

in the 2002 published *NetLino Internet Dictionary.* Until a few years ago, the Google search engine would not have found many entries for the term textiquette, and the only textiquette that existed did not contain any serious rules as example (87) proves.

(87) Do try not to resend the same message twice. It isn't cool and you don't want to look like a beginner, do you? (Anonymous 2003)

This, however, has changed: UPOC, North America's premier mobile community, formulated a more serious online textiquette in 2005 which includes 10 tips for texting. Overall, it can be concluded that the general netiquette manners apply and Grice's four maxims of conversation that are postulated in many netiquettes are valid to the point they make sense for text messages: one can, of course, not be a novelist, as the length of text messages is restricted, but rules such as not responding while being emotional or assuming that any sent message is permanent and might be seen by other people should be taken into consideration while texting. Similarly, Thurlow (2002) states that "the language of SMS appears to be driven by three key sociolinguistic 'maxims' […] all serving the principle of sociality which drives the messaging: brevity and speed, paralinguistic restitution and phonological approximation". The rule of not using capital letters also seems to be mandatory and is more closely defined in "Cap-Happy" (UPOC 2005), which declares that "a message in all caps is perceived as shouting, so unless you're really ticked off, or trying to emphasise a word, go with lowercase." In accordance with this rule, Crystal (2002: 136) speaks of the "lower-case default mentality" which means that "any use of capitalization is a strongly marked form of communication. Messages wholly in capitals are considered to be 'shouting', and usually avoided; words in capitals add extra emphasis". Many users, however, do not seem to know this rule, otherwise there would not be 16% of the text messages composed in all upper case letters (while 10% of the messages in the German and Swiss-German corpora are wholly capitalised). Nevertheless, some users are sensitive towards upper-case usage as the statement of the following sixth form student, Alice, shows: "I hate it when they do it [send text messages] in capitals […] ah, it annoys me" (Alice quoted from Taylor and Harper 2002: 443, my comment).

Another interesting rule, which, being rather descriptive than prescriptive, clearly expresses people's changed writing habits, is the one called "Spelling Schmelling" (UPOC 2005). It proclaims and officially allows spelling freedom when composing text messages: "This ain't grammar school. As long as it's clear, they'll get the message – no matter how you spell it." This rule reflects the changed attitude towards informal written discourse: text messaging is mainly intended for information exchange, and as long as the information can be put across, i.e. as long as the message is comprehensible, any chosen mode of writing is legitimate.

Besides prescriptive textiquettes, there are also pattern books and online SMS-collections which contain prefabricated text chunks that can be used to approach another person via SMS. These can include love-messages, birthday wishes, seasonal greetings as well as sayings for various other occasions, even ASCII pictures. Such ASCII pictures, as depicted in example (88), do not appear in emails but are characteristic of text messages. Not only pictures, but also sayings as the one in example (89) are exchanged, and it is often difficult to elicit whether such sayings are composed by the sender, quoted from some pattern books or simply forwarded. Hence, the author of the text message example (90) wonders if the sayings he received were actually made up by the sender or just quoted.

(88) >(. .)>
/ (v)
(,,) . . (,,). . .
Where are you?

(89) UR SO NICE UR SO KIND 1 IN A MILLION SO HARD TO FIND SO WHEN U GET DE MESSAGE THINK ITS TRUE THE 1 WHO SENT IT THINKS THE WORLD OF U XOXOXO

(90) Are you quoting all ties things or do you make them up? They are like music to my ears and very relevent take care

Pattern books, as they have come into fashion, can be compared to earlier letter manuals, and so, the SMS-boom of the years 2000 to 2003 has revived an old text type, namely the letter manuals as SMS-collection and manual. In contrast to traditional letter manuals, however, they now appear in a new look: electronically as Internet page, as CD-ROM or conventionally as SMS pattern-books (Ortner 2002: 207).

3.5 Features of the Involved Written but Conceptually Oral Language[6]

3.5.1 Salutation and Farewell

One of the major differences between text messages and other letter-like writing is the frequent lack of salutation. This is on a par with Androutsopoulos and Schmidt (2004: 63) who find that the most significant characteristic of salutations is their absence. Table 3-1 shows that 72% of the analysed messages do not contain a greeting[7].

Similar to emails, most salutations (24%) are *h*-salutations and the traditional *dear*-greeting only appears once in the whole corpus. The high percentage of omitted salutations indicates a certain *always-here-never-away* attitude towards one's friends or colleagues. Thurlow (2003) speaks of a "sense of perpetual contact" while Androutsopoulos and Schmidt (2004: 64) explain that the frequent lack of salutation formulas can be interpreted as a marker of dense SMS communication. SMS enables people to contact each other at almost any time around the clock. It is as if the other person is present even while being physically absent and as if he never actually leaves. Without leaving, however, people neither have to say goodbye nor hello and thus tend to approach each other without greeting formulas in their text messages. This again may lead to an equally high rate of omitted farewell formulas, which is indeed the case: the same percentage of text messages (72%) does not contain a farewell, as Table 3-2 shows. Some authors add crosses at the end of their text messages, but they cannot actually be considered leave-taking formulas as they symbolise kisses and are, if at all, rather to be categorised as emoticons, showing affection towards the addressee.

6 In this section, only single SMS will be analysed. A comparison of single SMS with linked ones follows in Chapter 4, Single Text Messages versus Linked Text Messages.

7 As in Section 2.5.1, Salutation and Farewell, I categorise the salutation and farewell formulas, thereby hazarding the consequences that the plurality of these formulas goes lost. However, the basic elements and tendencies can be best worked out this way.

Table 3-1: Salutations in Text Messages

Salutation categories	number	%
<first-/nickname>	10	1%
dear	1	0%
hello	12	1%
hey	125	14%
heya/hiya	17	2%
hi	62	7%
hy	2	0%
yo	10	1%
others	12	1%
none	656	72%
total	907	100%

Another reason for the lack of salutations and farewells may be due to the restricted length of text messages. Yet, considering the fact that the average SMS does not make use of the 160 characters available (cf. Section 1.3, Methodology: Data Collection and Corpora), this explanation seems inadequate, as a simple *hi* and *bye* would not use up more than seven characters at the most. What has to be taken into consideration, however, is the fact that many text messages in the corpus are part of a longer dialogue (thread) so that salutation and farewell formulas may be redundant. In this respect, one message could be equalled to a turn in a face-to-face or telephone conversation: turns in the middle of a conversation are not framed by *hi*'s and *bye*'s, nor are they in the middle of an SMS dialogue. Hence, it can be concluded that the lack of both salutation and farewell formulas is not primarily due to the restricted length of text messages, but due to the spoken-like, dialogic nature of this type of discourse as well as its quality of making the recipient seem present while actually being absent.

Table 3-2: Farewells in Text Messages

Farewell categories	number	%
<first name / initials>	57	6%
bye	5	1%
cheers	4	0%
good night	13	1%
(lots of) hugs	4	0%
later/s	5	1%
(lots of) love / love you	55	6%
see you <time adverbial>	34	4%
sweet dreams	2	0%
take care	5	1%
text back	45	5%
others	23	3%
none (of which xx)	655 (176)	72% (19%)
total	907	100%

3.5.2 Spelling

3.5.2.1 Misspellings and Typos

Text messages are even more ephemeral than emails: in contrast to emails, which can be stored in computer files, SMS cannot be saved for a long time as most mobile phones and SIM cards are highly limited in space. Important text messages might be copied into special booklets[8], but they cannot be saved as they are[9]. Hence, little attention is paid to

8 Kasesniemi and Rautiainen (2002: 178) recognise that it has become a habit of teenagers to "copy their messages into calendars, diaries or special notebooks designed for collection SMS messages. This practice of message collection is an important part of text messaging culture".

9 Meanwhile, there are certain programs available for (newer) mobile phones which make it possible to transfer one's text messages to the computer and save them there. These programs, however, are still not very widespread so that the majority of users

correct spelling. Also, there is no spell-check available for text messages, though it would not be of much use anyway: people write under real-time constraints, and as the typing of messages already takes up much time, there is no time left for correction. Especially in back-and-forth text messages, the composer of a message does not want to let his addressee wait and thus aims at an immediate response. Moreover, it is the information that matters: SMS is primarily an informative as well as expressive and phatic medium and so, content, not language, counts. Crystal (2001: 111), however, rightly states that "some misspellings can make the reader pause or make an utterance ambiguous or unintelligible". This indeed can lead to difficulties. Hence, it is not immediately obvious at the first reading that the word *treaky* in example (91) should be 'freaky' and that *dooze* in example (92) means 'booze'.

(91) That's so *treaky*! I was just thinking about you when my phone went. [Name] and I are at a ceilidh but we're going to be boring and leave soon. (My emphasis)

(92) *Dooze*.Check.fags.check.loud angry music.check OK, now you are ready 4 valentines! (My emphasis)

In fact, typos particularly make a difference at the beginning or at the end of a word whereas misspellings within a word are often not noticed[10].

either do not know about them or else do not have a mobile phone for which such programs exist at all.

10 There has been an ongoing discussion on the Internet about pattern recognition in language. This discussion, it is believed, has its origin in Saberi and Perrot's article about "Cognitive Restoration of Reversed Speech". Messages like "The paomnnehel pweor of the hmuan mind: Aoccdrnig to a rscheearch at smoe Uinervtisy, it deosn't mttaer in waht oredr the ltteers in a wrod are, the olny iprmoetnt tihng is that the frist and lsat ltteer be at the rghit pclae. The rset can be a total mses and you can still raed it wouthit porbelm. Tihs is bcuseae the huamn mnid deos not raed ervey lteter by istlef, but the wrod as a wlohe. Amzanig huh?" have been circulating for a while and Jim (2003) at Uncle Jazzbeau's Gallimaufrey's concludes that "Saberi's work sugsegts [that] we may have some pofrweul palrlael prsooscers at work. The resaon for this is suerly that idnetiyfing coentnt by paarllel prseocsing speeds up regnicoiton. We only need the first and last two letetrs to spot chganes in meniang". This explains why typos within a word matter less than at word boundaries. However, more accurate research on this topic is needed and an analysis whether word final *g*-clippings, for instance, contradict this theory would be an interesting topic of research (cf. also footnote 12 below).

In example (93), for instance, the jumbling of the letters /h/ and /t/ in the pronoun 'anything' is likely to be overlooked for two reasons: first, because it is a typo in the middle of the word and second, since 'anything' is already spelt informally (*g*-clipping), one might not immediately be aware of the typo, but rather recognise the informal spelling.

(93) GOOD LUCK WITH *ANYHTIN* YOU DO HONEY! NO DOUBT YOU WILL DO GREAT WHATEVER, OR WHERE EVER YOU ARE!! YOU DANCIN MUCH NOW CHICK? CAN'T BELIEVE IM GOIN BACK 2 NURSIN. (My emphasis)

Sometimes, it is difficult to determine whether a word has been intentionally misspelt or whether it is a typographical mistake. The spelling of the verb *gona* in example (94), for instance, could be a typo as well as an economic way of spelling *gonna*.

(94) HEY U,IVE GOT URE PHYSIOLOGY HANDBK,ILL BRING IT IN 2M,COS I AM *GONA* MAKE IT 2 LETCTURES NOW (My emphasis)

Misspellings, however, whether intentional or not, are a natural feature of text messages and "occur, regardless of the educational background of the writer, in any situation where there is fast typing and a lack of editorial revision" (Crystal 2001: 111). One is even excused when misspelling a word, for this is always attributed to the swiftness with which a message is composed and so "the reader [is not] going to make a social judgement about the writer's educational ability, on the basis of such data – a contrast with what *would* happen if someone wrote a traditional letter containing such errors" (Crystal 2001: 111).

3.5.2.2 Use of the Upper and Lower Cases

Another feature of economy is to write everything in upper or lower case letters. As pointed out above, capitalisation is equally classified as shouting in text messages as in emails and CMC in general (cf. Section 3.4, Rules and Conventions: Textiquette). Yet, many texters appear to be unaware of this rule and (still) regard messages that are all capitalised as legitimate. Hence, the analysed corpus contains more messages that are spelt exclusively with upper than with lower case letters and only around a third of all the messages use correct upper and lower case spelling, as Table 3-3 illustrates.

Table 3-3: Upper and Lower Case Spelling in Text Messages

Case	number of messages	%
all upper case	148	16%
all lower case	88	10%
correct upper and lower case	321	35%
inconsistent use of capitalisation/minusculisation	345	38%
half lower case, half upper case	5	1%
total	907	100%

Table 3-4: Inconsistent Use of Capitalisation/Minusculisation

Case	number of words	%
lower case instead of upper case	696	85%
upper case instead of lower case	33	4%
upper case for emphasis (words)	90	11%
total	819	100%

A few messages (1%) also mix the use of case and change the case mode in the midst of the written text as in quotation (95).

(95) hey there whats the plan 2nite? u can come n save me from a lads nite out?! IM SITTING IN VODKA BAR JUST NOW-[NAME] WANTS 2 SEE EVERY-ONE-YES HES A PSYCHO

All in all, 65% of the analysed messages deviate at some point from the traditional spelling rules regarding upper and lower case and 38% change case and either use lower case instead of upper case, or vice versa without being consistent. Table 3-4 shows that of those 819 words which are wrong in terms of case, 85% contain lower case instead of upper case letters, 4% upper case instead of lower case letters, and 11% of the words are capitalised for emphasis. It is noticeable that these results are very similar to the ones observed in the email corpus; they support the previous findings concerning the lax attitude and relative indifference towards the orthographical correctness of the written words in computer-mediated communication.

3.5.2.3 Non-Conventional Spelling

As in emails, shortenings and non-standard spellings are similarly characteristic features of text messages. However, their use is even more distinctive of SMS, as example (96) illustrates, where words are written together, abbreviated or contain letter and number homophones.

(96) CANICALLUL8ER?WHATTIMEDUFINISH2NITE?ILLTRY@7-ISH.BYE [NAME]X

Most of these features have already been dealt with in Section 2.5.2.3, Non-Conventional Spelling. What follows is thus a description of the results illustrated by some examples as well as some complementing remarks where necessary. Table 3-5 below summarises the spelling features that will be discussed.

Table 3-5: Non-Conventional Spelling in Text Messages

Spelling	number	per 1000 words
letter/number homophones	1237	68.48
g-clippings	203	11.24
t/*d*-clippings	16	0.89
consonant spelling	442	24.47
omission of apostrophes	312	17.27
emulated prosody / onomatopoeic exclamatory spelling	244	13.51
phonological approximation and other non-standard shortenings	678	37.53
total	3132	173.39

a) Letter and Number Homophones

Letter and number homophones are one of the predominant elements in text messages: they occur 68.48 times per 1000 words. The most widely used homophone is the letter homophone /u/; it appears 23.21 times per 1000 words and is followed by the number homophone *2*, which occurs 8.59 times per 1000 words. Homophones are used as a means of economy in order to save time and space. They are highly versatile as they can be combined with other homophones or lexical items. Where number homophones appear, words may be written together without spaces in between, for they visually divide the lexical entities as the text messages in examples (97) and (98) illustrate[11].

(97) I never feel that.I never dont wanna cu-theres no *need2nag* or think ull *have2cos* u dont.I just dont like it when im *trying2balance* everything.I love you xxxx (My emphasis)

(98) Hi,[Name]!How r u?I *want2meet* tomorrow ouside Clapham Junction st.at aprox.7 p.m.&what will be the best *time4u*?I finish work at6p.m. (My emphasis)

Homophones are not compatible with apostrophes. Whenever homophones are used in combination with a contracted verb form of an auxiliary or main verb, they are attached without an apostrophe. Quotations (99) to (101) provide examples of such conjoined words.

(99) No idea how *ull* do that! (My emphasis)

(100) *uve* cheered me up! (My emphasis)

(101) hey babe sounds cool. spoke 2 [name] this am. She's a bit down but OK. Can't wait 2 see everyone. I'll ring u later 2 find out plans. Hope *ur* having good day x (My emphasis)

Occasionally, homophones are also wrongly applied. The number homophone *4*, for instance, cannot be employed in the preposition *from* as in example (102), nor is the number homophone *9* in example (103) correct, for it does not constitute a proper homophone: in order to be able to read the word *f9* correctly as 'fine', one must delete the letter /n/ of *nine*. It then, however, no longer is a homophone on linguistic grounds.

---

11 For further illustrations on this aspect, see Section 9.2, New Trends – Old Features.

(102) Hi [name]! So hpy 2 hear *4m* u. Jus gt home *4m* coleg.we hav 2 giv 2 presntatns 2moz. spnt las lesn listnin 2 tcr in clas singn bob Marly an hmns. (My emphasis)

(103) HI, HOW R U? YE ITS *F9* IF WE MEET UP ON SAT ILL CALL U L8A TAKE IT EAZ! (My emphasis)

b) Clippings

G-clipping is a prevalent phenomenon in SMS and occurs 11.24 times per 1000 words; it affects the word final /g/ of nouns as well as verbs. Over a third of the *ing*-forms are clipped, i.e. 36%. Full *ing*-forms occur 21.27 times per 1000 words. There is only one instance where the clipped /g/ is substituted by an apostrophe, which is futile to the point that it does not serve economy nor is it time-saving in any way. Apostrophes are more difficult to type than any other letter of the alphabet and use up exactly as much space as the letter /g/. A clipped /g/ is therefore not usually marked by an apostrophe and the same applies for other clippings, too. Word final alveolar plosives, for instance, can also be dropped as in example (104) where the adverb *just* is spelt without /t/ and in example (105) where both the determiner *next* as well as the noun *weekend* lack their final alveolar plosives. It is important to notice, however, that the elision cannot be motivated by phonological factors, as in spoken language.

(104) I *jus* stopped by ur flat and u were out. If u wanna still go cinema ltr let me know. Im gonna sleep for a few hours. (My emphasis)

(105) Sis, I cant come *nex weeken* it will have2 be da week afta dat (My emphasis)

Similar to *g*-clippings, *t/d*-clippings constitute a phonological marker of register in so far as they mark the omission of word final alveolar plosives in pronunciation and thus further approximate the register of text messages to speech. Yet, they are relatively rare and merely appear 0.89 times per 1000 words. Their use depends on sociolinguistic aspects and so, people with an educated background like students might employ these features less often than people from other social classes. Or else, the rare use of *t/d*-clippings could as well be due to the fact that the register of text messages is still in flux so that features such as word final *t/d*-clippings will become more marked with the further development of the register. These hypotheses, however, cannot be verified at this point and further (sociolinguistic and long-term) research is needed to be able to make more accurate statements about these aspects.

c) Consonant Spelling

Consonant spelling is a feature which is characteristic of computer-mediated communication and quite frequently occurs in SMS: there are 24.47 instances of consonant spelling in the corpus. Taken in isolation, words whose vowels are deleted may be ambiguous and do not always have a particular significance. Hence, the word *wk* can have various meanings and refer to 'work', 'week', 'weak', etc. and so can the word *mst*, which may stand for 'most', 'must', 'mist', or 'mast'. However, these words are disambiguated in their specific context: thus *wk* in example (106) clearly means 'week'; *mst* in example (107) is 'must'.

(106) Few problems at home and I've gotta stick around which is unavoidable but was sposed to b going to [name] in 1*wk*. (My emphasis)

(107) Hey there sexy [Name]. im good thanx still training hard. what have u been up 2 recently? *mst* meet up sn when th rowin has fin. take care. (My emphasis)

The list of consonant spelling is long. There are words such as *dnt* for 'don't', *wknd* for 'weekend', *wnt* for 'want', *bt* for 'but', *gd* for 'good', *rly* for 'early', or *ws* for 'was'. Usually, these words are understood without any difficulties. More intricate are those lexical items which combine different spelling methods. The word *cn*, for instance, has two common meanings, either 'can' or, less obvious, 'seen'. Quotations (108) and (109) provide an example of each.

(108) *CN* U PICK ME+[NAME](HEDGEGILR)UP AT 8 2M PLZ? THNX XX XX (My emphasis)

(109) Hi u? Wht u bn up 2? I avnt *cn* him in ages. L8rs. B gd xxxx (My emphasis)

Whereas the word *cn* meaning 'can' simply constitutes an instance of consonant spelling, *cn* meaning 'seen' combines two spelling methods: consonant spelling and the use of the *c*-homophone, meaning 'see'. Similarly, the word *wv* meaning 'with' (110) combines consonant spelling with *th*-fronting, i.e. the replacement of the RP dental fricative [ð] by the labiodental fricative [v], which is quite common in the London area and, in fact, a stereotype of London English. The same explanation holds for the word *fnk* meaning 'think' (111): here, in addition to consonant spelling, the RP dental fricative [θ] is replaced by the labiodental fricative [f].

(110) Hey [name] soz mums bin stresd+ws lyk"il do it l8a"bout the bkin fee bt she neva did lol.bt let me no hw it goes+if its gd il try go *wv* u nxt tym? (My emphasis)
(111) hey, hope yarght,wha yup to? Jus watchin they *fnk* its al ova,funny as fook lol,I knw iv bin a pain l8ly jus bare wiv me=Dplz x x tb if can x x (My emphasis)

As pointed out in Section 2.5.2.3, Non-Conventional Spelling, liquids might as well be affected by consonant spelling. Yet, it is not only the liquid /r/ but also the liquid /l/ which tends to be deleted. What is more, liquid deletions do not merely appear in consonant clusters, but also in pseudo-phonetic spelling. Accordingly, the liquid /l/ in the modal verb 'would' is deleted in both the consonant cluster *wdnt* (112) as well as in the pseudo-phonetic spelling *wud* (113).

(112) [name] [name] & [name] have came in so drunk u *wdnt* believe it, rolling in the grass & everything, shocking! (My emphasis)
(113) NO!He he!That *wud* b awful.XXX (My emphasis)

Also affected by consonant spelling is the letter /c/ in combination with /k/. Since the consonant /c/ does not represent a distinct phoneme and does not serve any purposes with respect to pronunciation or meaning, it can easily be deleted when preceding the consonant /k/. Hence, the adverb 'back', for instance, is spelt *bk* in consonant clusters or *bak* in pseudo-phonetic spelling, as examples (114) and (115) illustrate.

(114) thought u'd come out wi us ON TUE BUT U AWAY.TXT WHEN COME *BK*. [NAME] TXT *BK* (My emphasis)
(115) Will u be *bak* b4 10? (My emphasis)

Another interesting aspect is the *ing*-ending in consonant clusters. The verb 'texting' in example (116) could either be spelt *txtng* or *txtn* if the word final *g*-consonant is also clipped. However, dissonant with the principle of consonant clusters to lose all vowels, it is spelt *txting*.

(116) Jst *txting* t say hello. (My emphasis)

This observation supports the assumption in Section 2.5.2.1, Misspellings, Typos and Transfer Errors, that consonant spelling simply affects the stem of the word and not its ending. Yet, there are not enough examples in the corpus to actually test this hypothesis, but *ing*-endings nevertheless tend not to be affected by consonant cluster simplifications. There is no example but one in the corpus that contradicts this finding (cf. quotation 1)[12]. Quotations (117) to (119) provide further examples in support of this hypothesis.

(117) Your DC is crass. Its scrtched & keeps *skpin* trax Send me anover plz. Tanzxxx (My emphasis)

(118) South France. I was *jkin*. I mite gt u sumat from france if ur lucky. Wat did u gt me huh? Hu u goin Scotland wit? X (My emphasis)

(119) [Name] did e! She bare dumb. Shes bre spaced+fukdup. Shes *cmin* 2 th gig y tambien [name]. Hv u eva rly met [name]? Tbx hw ru? X (My emphasis)

d) Apostrophes

In contrast to the email corpus, in which only 18% of the apostrophes are omitted, almost half of all the apostrophes (43%) are deleted in text messages: apostrophe deletions occur 17.27 times per 1000 words. This high percentage is mainly due to the typing effort: keying in an apostrophe on a touchpad of a mobile phone requires a significantly higher typing effort than on a computer keyboard. Thus not only contracted forms (as in emails), but also possessives lack their apostrophes as depicted in examples (120) to (122)[13]. What is more, there is no single instance of apostrophe misuse: no plural form, no plural acronym or initialism is written with an apostrophe.

---

12 As pointed out above in footnote 10, we only need the first and last two letters to spot changes in the meaning of words. In verb forms ending in *ing*, the final /g/ may be clipped. It has been shown that this constitutes a phonological marker of register and is further the result of immediacy, i.e. of brevity and speed. To still perceive a given verb in its continuous form, the last two letters of the *in[g]*-ending, i.e. the /in/, must thus remain. This is one explanation towards the fact why the vowel /i/ in continuous verb forms is not affected by consonant spelling.

13 For reasons of clearness, names are not substituted by the common placeholder [name] in these text messages. Yet, to still guarantee anonymity, pseudonyms are applied.

(120) Hey my lovely! Am at *euans* so wont b comin out 2nite! Have fun thou! C u 2moro x x x x (My emphasis)

(121) Aahh gr8 film!Do u not speak 2 [name] much at skool then? Dont wanna go back on thurs! Wen r u gonna get a job lazy-scar? Do u live most at ur *dads* or *mums*?xx (My emphasis)

(122) HEY GIRLY U GOIN *HELENS* ON FRI? WANNA BUS IT 2 MYN B4HAND? SAFE X (My emphasis)

e) Emulated Prosody and Onomatopoeic Exclamatory Spelling

Text messages are very simple in their format and layout and there is no possibility of underlining single words, using italics, bold letters or colours. Yet, this is not a new problem of computer-mediated communication and so, the traditional "use of graphological means to emulate suprasegmental phonological features, such as emphasis and stress" (Bergs 2003: 4) has been taken over from email and all other computer-mediated communication. These graphological means include reduplication of letters and words, capitalisation, asterisks as well as the creative use of punctuation and other "textual representation of auditory information such as [...] laughter and other non-language sounds" (Herring 2001: 617). All these practices result in unconventional orthography as illustrated in examples (123) to (127). Herring (2001: 617) concludes that these strategies, "rather than reflecting impoverished or simplified communication, demonstrate the ability of users to adapt the computer medium to their expressive needs", and she recognises that "this results in a linguistic variety that, despite being produced by written-like means, frequently contains features of orality".

(123) *Awww* my darling i love u so much! (My emphasis)

(124) Ah [name], u know i wanna b in a music video? Omg, i can b in the rifles video and get *FREE DRINK*! Xxx (My emphasis)

(125) I *looove* you*!!!!! HeHe.....* (My emphasis)

(126) Here's a big snog for you. *Slurp slurp mmmwwahh*! (My emphasis)

(127) 85% chance of [name] getting awesome unix job. Good place/money/career and general answer to prayer. **bounce**bounce** *:-) *Yay Yay Yay* (My emphasis)

Emulated prosody as well as onomatopoeic exclamatory spelling are prevalent features in text messages: they appear 13.51 times per 1000 words. The huge creativity in messages employing emulated prosody gives them a very personal touch, so that much of the coldness of the medium is made up for.

f) Phonological Approximation and Other Non-Standard Shortenings

In terms of the sociolinguistic maxims of paralinguistic restitution and phonological approximation, it can be noted that the former "seeks to redress the apparent loss of [...] socio-emotional or prosodic features as stress and intonation", whereas the latter "adds to paralinguistic restitution and engenders the kind of playful, informal register appropriate to the relational orientation of text-messaging" (Thurlow 2002). Thereby, texting has adopted ludic qualities and allows the user to play and experiment with the language, its words and rules. Instances of phonological approximation and other non-standard shortenings occur 37.53 times per 1000 words. Table 3-6 provides an overview of the most common spelling tendencies in the analysed text messages and formulates seventeen rules which are relevant for the composition of text messages. None of these rules, however, applies to the composition of emails[14]. This set of rules can thus be regarded as a marker of SMS – constituting a major difference to emails.

The most common feature in text messages to approximate the phonological value of specific sounds is the replacement of /th/ by the letter /d/ in words like *dis*, *den* or *de*. This feature occurs 4.88 times per 1000 words. If /th/ is replaced by /d/ before a word final [ə], it is often combined with another rule, namely the rule to replace a word final [ə] by /a/ or the rule to vocalise word final /r/ respectively. This rule is quite common, too: there are 3.71 instances of such *a*-endings per 1000 words. Examples of dental fricatives turning into alveolar plosives as well as examples of word final [ə] being replaced by /a/ and word final *r*-vocalisations are provided in quotations (128) to (131).

14 The only rule which shows a certain tendency in the email corpus is the *should/u*-rule: it occurs 0.44 times per 1000 words, followed by the *night/ite*-rule which occurs 0.18 times per 1000 words. All other rules are either not present at all or applied rarely so that they cannot be regarded a typical feature of emails and therefore need no further discussion. For more information, cf. Section 5.2, Spelling Tendencies.

Table 3-6: Spelling Tendencies in Text Messages

Spelling rules	example	number	per 1000 words
again [e] → /e/	again → agen	10	0.55
any [eni] → /ne/	any → ne	8	0.44
dental fricative [ð] → labiodental fricative [v] → /v/	with → wiv	12	0.66
dental fricative [ð] → alveolar plosive [d] → /d/	the → de	88	4.88
dental fricative [θ] → labiodental fricative [f] → /f/	think → fink	17	0.94
double consonant → single consonant	call → cal	48	2.66
double vowel → single vowel with the corresponding pseudo-phonetic value	been → bin	15	0.83
dropping of word initial /h/	having → aving	11	0.61
intervocalic /r/ → word final /z/	tomorrow → tomoz	21	1.16
night [aɪt] → /ite/	night → nite	50	2.77
should [ʊ] → /u/	should → shud	16	0.89
some [ʌ] → /u/ (+ dropping of word final /e/)	some → sum	70	3.88
though [əʊ] → /o/	though → tho	40	2.22
time [aɪ] → /y/ (+ dropping of word final /e/)	time → tym	26	1.44
word final /s/ → word final /z/	hols → holz	22	1.22
word final /y/ → /i/	already → alreadi	5	0.28
word final [ə] → /a/ / vocalisation of word final /r/ → /a/	the → da / never → neva	67	3.71
total		526	29.14

(128) Aww thnku [name] ur *da* bomb!&thnkx so much4*da* card absolutli luvli!luv u!xxx (My emphasis)

(129) Hey,[name],lol,u aite,howz ur holz goin? Tb! Luv [name] 4rom skol,*dis* my *oda numba*,keep *da oda* 1 aswel! XxX (My emphasis)

(130) Y r u bein lyk *dis* [name]!! *ders* rele no need even [name] finks it perfetic bt its kool ur obv stress evn tho ur kinda in *da* rong bt *wateva* i dnt… (My emphasis)

(131) Alr.texu.Em wel we goin 4 drinks *afta* wrk tomoz. Nt sure about 2nite…[Name] sed its still on.i'll go tomoz.havea good day.xxxx (My emphasis)

Another frequent feature of pseudo-phonetic spelling is the replacement of /o/ by /u/ where pronounced as [ʌ] in words such as *luv* (132), *sum* (133), *cum* (132, 134), and *dun* (135). In the wake of pseudo-phonetic spelling and economy, these words may further drop their word final /e/.

(132) HEY IS IT K IF I *CUM* C U l8r?I NO DAT U WERENT FEELIN SO GD EARLIER.HOPE U R FEELIN BETTR SOON.*LUV* S X (My emphasis)

(133) ive almost finished work… ru at home?I need to get *sum* credit + then ilb rnd2 urs.cu soon x (My emphasis)

(134) R U OK.WAT U UP 2 2NITE.IF U R NT DOIN ANYTHING THEN *CUM* ROUND MINE (My emphasis)

(135) WAT?!?! u didnt gt it *dun*?!?!Ahhh wts she gna say? She gna b pissed! (My emphasis)

An interesting though unusual feature is the spelling of [aɪt] as /ite/ as in *nite* (meaning 'night'), *rite* (meaning 'right'), and *mite* (meaning 'might'). On the one hand, the letters /g/ and /h/ are reduced, but on the other hand, a word final letter /e/ is added. This is a rather unusual procedure considering that the general spelling tendency points towards spelling economy so that the additional letter /e/ at the end of the word becomes a highly questionable phenomenon. Interestingly, the letter /e/ is not added where [aɪt] is spelt /yt/ as examples (136) to (138) illustrate. This way of spelling [aɪt], however, is much less frequent: it occurs 0.44 times per 1000 words, which is around a sixth as often.

(136) *Gdnyt*, Love always, me xxx (My emphasis)

(137) Lifes gud i spose.Can i ring u *2nyt* cos i wanna talk 2 sum1 bout something? *Myt* as well b u! Hows things in the big city?x x x (My emphasis)

(138) Sori babe its just sumfin dat apend b4 im just pisd of. U *alryt*? TB Luvyaxxxxx (My emphasis)

Furthermore, there is a tendency to reduce double consonants to single consonants as in *wel* and *hapy*. This becomes more complex if combined with an omitted apostrophe as illustrated in example (139), or if occurring in combination with consonant spelling plus omission of apostrophe as in example (140).

(139) Chicken is it ok 4 me 2 bring sum1else tmw?She mite not cum but she doesnt want 2 cum unless its ok way u.I said *itl* prob b ok way u.*Il* b up abt 8 ok. (My emphasis)

(140) ok- *wl* do... (My emphasis)

Slightly less frequent but still prevalent is the spelling of [əʊ] as /o/ as in 'know' or 'though' in examples (141) and (142) or the spelling of [aɪ] as /y/ as in 'time', 'like' or 'mine' in quotations (143) and (144). Again, the word final /e/ of these words is usually dropped, similar to the case of the *some/u*-rule described above.

(141) this may be a little short notice but i'm havin some ppl round on monday4my birthday.wondered if u'd like2come.let me *kno* yea.lovya,[name].x (My emphasis)

(142) Y r u bein lyk dis [name]!! ders rele no need even [name] finks it perfetic bt its kool ur obv stress evn *tho* ur kinda in da rong bt wateva i dnt... (My emphasis)

(143) lol! i wnt slp registration. darling wot *tym* did u go slp yday i was *lyk* i cnt memba... txt me l8a. tb. Xxx (My emphasis)

(144) HEY GIRLY U GOIN [NAMES] ON FRI? WANNA BUS IT 2 *MYN* B4HAND? SAFE (My emphasis)

Another common characteristic in text messages to approximate the phonological value of specific sounds is the substitution of the word final /s/ by /z/ in words such as *dayz, howz,* or *sparklerz*. Related to this rule but nevertheless of a completely different nature is the rather striking feature of the intervocalic /r/ turning into a word final /z/[15] when the last

15 This, in fact, constitutes a remarkable feature and reminds of the observed /s/ – /r/ alternation in Pre-Latin. Trask (1996: 78 f.) explains that "Pre-Latin had the phonemes /s/ (a voiceless sibilant) and /r/ (a voiced liquid). At some stage, /s/ became voiced to [z] between vowels (a common assimilation)" which then "was lenited to a liquid [r]. In doing so, it became identical to the existing /r/, and hence all words formerly containing intervocalic /s/ now had intervocalic /r/ instead". An example is given by the Latin word *flōs* (meaning 'flower') which formerly had the perfectly

syllable of words such as ‘sorry’ and ‘tomorrow’ is dropped. The word ‘sorry’ then becomes *soz* and the adverb ‘tomorrow’ turns into *tomoz*, as example (145) shows.

(145) Hey [name], *soz* bout eng 2day, i didn’t seem wiv it did i? Im *soz*, neways u wanna meet up *2moz* p.3 or 4 *2moz*? Tb, xOx (My emphasis)

Other pseudo-phonetic spellings reflect regional dialects as the *th*-fronting, i.e. the replacement of the dental fricatives [θ] and [ð] by the labiodental fricatives [f] and [v] in words like ‘think’ (146), ‘with’ (147), ‘nothing’ (147), and ‘thing’ (147). This is a feature usually associated with Cockney speech, but currently, it is “spreading regionally from London to other areas where it has not been observed before” (Altendorf 2003: 63).

(146) Yo [Name] baby! r u @ yr flat? *fink* i got class today but pop by @ 2:30 (My emphasis)
(147) I love you so much! I want to b *wiv* you still, Im sorry 4 wot I did, she means *nofin* to me, I swear! Take me bk, u sexy *fing*!xxxx! (My emphasis)

Another well-established feature in the history of English is *h*-dropping: it is “one of the most widespread features of non-standard urban English accents” in England and – along with *th*-fronting – another stereotype of London English (Altendorf 2003: 62). The replacement of RP /h/ by zero in word initial position is a recognised element also in text messages as examples (148) to (150) show.

(148) Sori babe its just sumfin dat *apend* b4 im just pisd of. U alryt) TB Luvyaxxxxx (My emphasis)
(149) I would *av* eaten a whole jar of mint sauce (My emphasis)
(150) Hey, babez, hope ur *avin* a gd tym and glad 2 b bak at graveney. c ya lates , xx (My emphasis)

---

regular plural form *flōses* but then came to have the irregular plural *flōres*. I.e. the stem acquired an alternation between the forms *flōs-* and *flōr-* depending on whether it is followed by a vowel or not. What now is observed is a reverse order of this development: intervocalic /r/ turns into /z/ if the ensuing vowel and syllable respectively is dropped. The reason why /r/ rather turns into /z/ than into /s/ is explained by the rule, i.e. by the tendency, to spell a word final /s/ as /z/ in text messages.

Examples (151) to (153) illustrate another kind of pseudo-phonetic spelling, namely instances of the *should/u*-rule. According to this rule, the [ʊ] in words like 'should', 'would', and 'could' is spelt as /u/.

(151) Sorry was in the shower then got scared by phsyco n jumped out! Shes great as ever. Ill tell her u asked. ye xmas is fine of course. battersea park. *Shud* b go. tbx x (My emphasis)
(152) Yo Yo smelly girlwhat time *wud* you like us 2b at your house later? & wot time do we have to leave? luv ya [Name] xxxx (My emphasis)
(153) *Cud* u get sir 2 sign me in PlZZZ? [NAME] X (My emphasis)

Occasionally, double vowels as in 'been' or 'good' appear to be reduced to a single vowel with the corresponding pseudo-phonetic value as quotations (154) and (155) show.

(154) Wats goin on b.c how u stay if I dnt fne u,u dnt fne me.But u *bin* alrite [name]? (My emphasis)
(155) Hi babe,had a *gud* wkend? (My emphasis)

A further interesting feature is the spelling of 'any' as *ne* where the letter /n/ stands for the [en]-sound and the letter /e/ for the [i]-sound of [eni]. This spelling variant also occurs in compounds as examples (156) and (157) prove, *ne1* meaning 'anyone' and *ne mre* standing for 'anymore'[16]. Similarly, the letter /m/ in example (158), meaning 'them', comprises the [em]-sound. In spoken English, the /th/ in the pronoun 'them' can fall together with the word final /t/ of the verb 'get' and is then pronounced as [getem] or [geʔem] so that that the letter /m/ is sufficient to pseudo-phonetically spell the word 'them' in the given example.

16 Surprisingly, the adverb 'anymore' is spelt as two words. The only reason for this phenomenon is the legibility of this highly abbreviated word. Accordingly, *ne mre* is easier to read than *nemre*. Economy wise, this way of spelling 'anymore' does not contribute much to efficiency and economy: all in all, there is only one character saved. Instead of the seven characters in 'anymore', *ne mre* uses up six. This example shows that it is not only a matter of efficiency or economy which leads to such inofficial spelling variants, but also a question of code. Texters have started to use their own code irrespective of its benefit.

(156) Hey do you kno *ne1* at uni in the eastendish tht wud b interested in livin out coz I found a place 4/5 bedrooms £440 a week.so lemme kno Tx (My emphasis)
(157) Hey,hw u doin? I wnt 2gt it dun,bt nt2sure *ne mre.* wt ufink i shud do? Xxx (My emphasis)
(158) HEY [NAME], NOT TOO BOTHERED BOUT A PRESSIE BUT IF UR STILL WANDRIN ROUND CAM SUM BRACES WUD B KL, U CAN *GET M* IN NE VINTAGY SHOPS. LUV U, [NAME], XXX (My emphasis)

Finally, the word final /y/ in words like 'already', 'happy', or 'sorry' may be spelt /i/ as in examples (159) and (160) or in example (148) above, while the [e]-sound in words like 'again' and 'said' appears to be spelt /e/ as shown in examples (161) and (162).

(159) [Name] [name] and [name]..me readin asian woman.lol.haha i tld da man i *alredi* gave him my card. (My emphasis)
(160) Hey again. Do you think i can call you on sunday again? I miss talking to you hehe. You ARE very very special (and i'm drunk and *hapi* :-)) lotsa hugs!! (My emphasis)
(161) Hello Sexy sorry i aint rung u no creds u looked buf on sat if u want 2 cum out and c me *agen* txt BK.x (My emphasis)
(162) Alr.texu.Em wel we goin 4 drinks afta wrk tomoz. Nt sure about 2nite…[Name] *sed* its still on.i'll go tomoz.havea good day.xxxx (My emphasis)

These rules clearly show that orthography in text messages is not as random and unstable as one might think. There are, in fact, underlying rules that make the messages still comprehensible. Nevertheless, pseudo-phonetic and user-idiosyncratic orthography as well as ad-hoc abbreviations can lead to a variety of different spellings of the same word. Hence, there are eight different non-standard spelling variants of the word 'tomorrow' in the corpus, namely *2m*, *2mz, 2moro*, *2morrow*, *2moz*, *tomm*, *tomoz*, and *tmw*. Similarly, the word 'tonight' is spelt *2ngt*, *2nite, 2nyt*, or *tnight*, while the pronoun 'you' appears as *y*, *ya*, *ye* and *u*. Besides, abbreviated forms can also result in homonyms. The letter /t/, for instance, either stands for 'the' or 'to', whereas the letter /d/ can refer to the definite article 'the' or the verb 'do', and it is only in context that these shortenings become disambiguated. These findings show that the use of shortenings and pseudo-phonetic spelling in text messages is also a matter of spontaneity and creativity and allows a certain spelling freedom.

3.5.3 Punctuation

As has been pointed out in Section 2.5.2.3, Non-Conventional Spelling, "for the sake of paralinguistic restitution [...] multiple punctuation may be more desirable [...] [and] becomes more semantically marked" (Thurlow 2002). It can endow emphasis and intensify the written word depending on the chosen punctuation marks and their multiplicity. Example (163a) employs multiple punctuation marks and thus evokes a rather dramatic effect; the same message would be much less concise and effective with single punctuation marks as example (163b) illustrates.

(163a) oi!!! i'mlonely, whhere are u...?
b) oi! i'mlonely, whhere are u?

Yet, due to brevity and speed, punctuation marks can also be neglected. According to Table 3-7, essential punctuation marks are omitted 20.22 times per 1000 words.

Table 3-7: Punctuation Marks in Text Messages

Punctuation marks	number	per 1000 words
dashes instead of full stops, commas, question or exclamation marks	27	1.50
inappropriate punctuation marks	98	5.43
omission of essential punctuation marks	365	20.22
question marks and exclamation marks combined	17	0.94
trailing dots (full stops)	65	3.60
trailing exclamation marks	30	1.66
trailing question marks	2	0.11
single exclamation marks	428	23.69
single full stops	859	47.55
single question marks	453	25.10

The omission of essential punctuation marks often results in asyndetic, non-coherent and shopping-list like sentences that merely string information together without structuring it, as this is the case in example (164).

(164) Hi! was thinking about you earlier going to [names] tonight feel free to come down later? Had a great time in france will have to catch up !

Exclamation and question marks compete against full stops: there are only around twice as many full stops (51.15 instances per 1000 words) as exclamation (25.35 instances per 1000 words) and question marks (25.21 instances per 1000 words). The frequent use of exclamation and question marks is certainly due to the dialogic character of text messages. Questions concerning the recipient's well-being, arrangements and phatic information abound in SMS, and for the sake of emphasis, simple statements are often marked by exclamation points. Also, it is predominantly full stops and not exclamation or question marks that are omitted, as these are rather neutral and, as long as they do not appear as trailing dots, do not add extra emphasis to the message and can thus easily be neglected. It is therefore only full stops that are omitted in example (164). Example (165) further illustrates the frequent use of exclamation marks: apart from sentences (4) and (5), every other sentence could as well be marked by a full stop, which would have been equally appropriate but certainly not as expressive and emphatic.

(165) (1) Thanks [name], u did better than me!! (2) its my essay title! (3) I have 2 discuss that! (4) Aaah! (5) Thank you-lifesaver!

In tune with the hypothesis in Section 2.5.3, Punctuation, full stops tend to be deleted where emoticons appear as quotation (166) proves, whereas question and exclamation marks remain and are combined with emoticons as examples (167) and (168) show.

(166) I'm a total wreck again :-(can't stop crying. I've never been this sad and hurt before in my life :-(i don't know what to do/think. Life sucks :-(

(167) Must tidy flat!:(

(168) What ye up to later?:-)

Full stops can further be substituted by dashes, as example (169) illustrates. According to Ferrara et al. (1991: 20), dashes are a characteristic feature of computer-mediated language in general; they "represent the less clearly defined sentence-endings that are often the norm in conversation" (Maynor 1994: 50). As an almost equally common feature as in emails, they occur 1.50 times per 1000 words.

(169) IM SITTING IN VODKA BAR JUST NOW–[NAME] WANTS 2 SEE EVERYONE–YES HES A PSYCHO

3.5.4 Grammatical Features

3.5.4.1 Syntactic Reductions

Syntactic reductions are 1.6 times as frequent in text messages as in emails: according to Table 3-8, they occur 58.85 times per 1000 words. Döring (2002: 14) suggests that the high amount of syntactic ellipses in text messages surpasses the rate of word-deletions not only in other types of written communication, but also in oral communication.

Table 3-8: Elliptic Phrases in Text Messages

Ellipses	number	per 1000 words
articles / possessive pronouns	84	4.65
auxiliary verbs (without subject deletion)	133	7.36
be as a main verb (copula deletion) (of which existential *there*/article + copula deletion)	104 (18)	5.76 (1.00)
conjunctions	201	11.13
infinitive marker *to*	3	0.17
prepositions	30	1.66
pronouns other than subject pronouns	17	0.94
subject pronouns	386	21.37
subject pronouns and auxiliary verbs	105	5.81
total	1063	58.85

There can be multiple ellipses within single sentences, which results in a rather fragmentary style as shown in quotation (170) where there are several kinds of syntactic reductions, namely article, copula, subject, and pronoun deletions.

(170) TUT TUT wot…? U KNOW THAT I'VE ON Ø MO ON MY FONE.WOT Ø U UP 2NOW DAYS? ø thought u'd come out wi us ON TUE BUT U Ø AWAY. TXT Ø WHEN Ø COME BK. [NAME] TXT BK

The most common syntactic reduction concerns the interactive pronouns *I* (412 *I*-subject ellipses) and, considerably less frequently, the interactive pronoun *you* (56 *you*-subject ellipses), whereas the remaining 40 deletions either affect the *it*-, *we*-, or existential *there*-subjects. The high number of *I*-deletions is due to the private, interpersonal context in which texting takes place (Döring 2002: 14) and "correlates with the claim that SMS communication is generally an informal, spoken and proximity-orientated variety of language, as 1st and 2nd person pronouns also figure, for example, in Biber's involved dimension of register variation" (Bergs 2003: 7).

Syntactic reductions also affect auxiliary verbs, *be* as a main verb, the infinitive marker *to*, prepositions as well as conjunctions. Auxiliary verbs are even more frequently omitted (7.36 times per 1000 words) than *be* as a main verb (5.76 times per 1000 words), which marks a significant difference to the email corpus in which copula deletion is more dominant: whereas auxiliary verb deletions occur 1.3 times more often than copula deletions in the SMS corpus, *be* as a main verb is 4.1 times more often deleted than the almost never absent auxiliary verbs in the email corpus. This finding clearly proves that the tendency is towards an even more reduced sentence style in text messages compared to emails.

Auxiliary verbs further appear to be deleted in combination with subject pronouns (5.81 times per 1000 words), as shown in examples (171) and (172). In contrast to auxiliary verb and copula deletions, the deletion of the infinitive marker *to* is not a very common feature. It only occurs 0.17 times per 1000 words, whereas preposition deletions are slightly more prevalent, appearing 1.66 times per 1000 words.

(171) øø just chilling at the mo with a bottle of wine and some friends.
(172) Hello you :-) øø wanna chat sometime? I can call you anytime if you like. How's life going? Tried to call you a minute ago. Hope things are well. Hugs -me

Considerably more common, in fact the second most frequent syntactic reduction following subject deletion in rank, is the ellipsis of conjunctions. This results in asyndetic sentences of the kind illustrated in quotations (173) and (174).

(173) Going to sleep now will can [= call] you in morn. (My comment)
(174) HEY,TOTALLY MAJOR COLD, FEEL REALY CRAP SO I AINT COMIN OUT, SORRY GUYS. ITS WAS G8 2 C U BOTH 2DAY,IL C U AGAIN SUM TIME IM THROO. SORRY X

According to Table 3-9, asyndetic sentences occur 11.13 times per 1000 words, contributing to the fragmented style of text messages and computer-mediated communication in general.

Table 3-9: Conjunctions in Text Messages

Conjunctions	number	per 1000 words
asyndetic	201	11.13
coordinating	243	13.45
subordinating	169	9.36

Asyndetic, coordinating and subordinating sentences balance out one another; they appear between 9.36 and 13.45 times per 1000 words. Table 3-10 shows that the coordinating conjunction *and* is used most often (7.81 instances per 1000 words), followed by the coordinating conjunction *but* (3.77 instances per 1000 words) and the conditional subordination *if* (3.60 instances per 1000 words). For a closer analysis of these conjunctions, consider Section 2.5.4.1, Syntactic Reductions.

Table 3-10: The Most Frequent Conjunctions in Text Messages

Conjunctions	number	per 1000 words
coordinating conjunction *and*	141	7.81
coordinating conjunction *but*	68	3.77
other coordinating conjunctions	34	1.88
causative subordination	33	1.83
conditional subordination (of which *if*-conjunction)	67 (65)	3.71 (3.60)
other subordinating conjunctions	65	3.60

3.5.4.2 Verbs

The results of verbs in SMS are almost identical with those in the email corpus. The only deviation concerns activity and mental verbs: whereas there are slightly more activity than mental verbs in the email corpus, the relation is reversed in the SMS corpus. Nevertheless, together with the copula *be*, they still belong to the three most frequently used verbs, as can be inferred from Table 3-11 and Table 3-12. What is more, activity

Table 3-11: Verbs in Text Messages

Verbs	number	% of verbs	per 1000 words
activity verbs	368	12%	20.37
be as a main verb	747	24%	41.35
communication verbs	95	3%	5.26
mental verbs	520	16%	28.78
obligation/necessity modals	41	1%	2.27
permission/possibility/ability modals	128	4%	7.09
volition/prediction modals	247	8%	13.67
pro-verbs (e.g. *do, get, put*)	197	6%	10.91
others	822	26%	45.50
total	3165	100%	175.20

verbs occur "overall [...] much more commonly than verbs from any other semantic domain" (Biber et al. 1999: 265), i.e. they are not only a typical feature of speech, but also of writing and are thus not a distinctive element of conceptual orality. The difference can therefore be neglected. The only difference which may be a slight indication of SMS being closer to conceptual orality than email is the fact that mental verbs occur 1.28 times as often in the SMS corpus as in the email corpus. Just as in the email corpus, pro-verbs as well as volition/prediction modals are also rather common, appearing 10.91 and 13.67 times per 1000 words. They follow *be* as a main verb and activity/mental verbs in rank and constitute further characteristic features of conceptual orality.

Table 3-12: Modal Verbs in Text Messages

Modals	number	% of modals	per 1000 words
permission/possibility/ability			
can	94	23%	5.20
could	15	4%	0.83
may	4	1%	0.22
might	15	4%	0.83
obligation/necessity			
be supposed to	2	0%	0.11
have to	15	4%	0.83
must	3	1%	0.17
need to	9	2%	0.50
should	12	3%	0.66
volition/prediction			
be going to	56	13%	3.10
shall	2	0%	0.11
will	153	37%	8.41
would	37	9%	2.05
total	417	100%	23.08

3.5.5 Lexical Features

3.5.5.1 Prefatory Expressions

Prefatory expressions are a major feature of spoken language. They occur more often in text messages than in emails, which again indicates that SMS is more closely related to conversation. According to Table 3-13, interjections are the most frequently used prefatory expressions (appearing 9.47 times per 1000 words), followed by response forms (5.10 instances per 1000 words) and the linking adverb *so* (2.77 instances per 1000 words).

Table 3-13: Prefatory Expressions in Text Messages

Prefatory expressions	number	per 1000 words
discourse markers	14	0.78
interjections	171	9.47
linking adverb *so*	50	2.77
response forms	92	5.10
stance adverbs	13	0.72
total	340	18.84

Whereas response forms are the rarest feature in the email corpus, they belong to the two most common types of prefatory expressions in SMS, which implies that texting involves frequent back-and-forth dialogues. Half of all the prefatory expressions used, however, are interjections. Interjections, as Biber et al. (1999: 1083) point out, express the speaker's emotion. They are generally used very spontaneously and thus indicate that texting can be a matter of near real-time discourse, approximating the synchronicity of face-to-face or telephone conversation.

3.5.5.2 Spoken Vocabulary

As in the case of prefatory expressions, spoken vocabulary is more common in text messages than in emails. Table 3-14 shows that it occurs 24.87 times per 1000 words, which is more than twice as often as in the email corpus.

Table 3-14: Spoken Vocabulary in Text Messages

Spoken vocabulary	number	per 1000 words
regionalisms	13	0.72
slang expressions	113	6.26
spoken words/structures (excluding prefatory expressions)	307	17.00
taboo expletives (excluding moderated expletives)	16	0.89
total	449	24.87

Spoken vocabulary comprises words such as *bout, ya, baby* or any other oral structures, as example (175) indicates in which the sentence adverb *though* is used at the end of the clause to relativise the previous statement. Similarly, the collocation *none of my business* in example (176) or the use of the adjective *pretty* in example (177) constitute typical spoken expressions.

(175) I miss you.I love you much more *though.* (My emphasis)
(176) [Name] was saying he saw u it's *none of my business* and im sorry I was a bitch!;o) (My emphasis)
(177) It's *pretty* hard getting thru to my dad. I've left a message with my sister but don't know sorry (My emphasis)

There are also a few regionalisms such as *bonnie, aye, fit like, lass*, and *wee*, which are characteristic expressions of the English dialect spoken in the north of England and Scotland. Besides, there are slang expressions such as the adjective *pants* in example (178), which became a slang word in the 1990s, meaning 'nonsense', 'rubbish', 'bad'. Both regionalisms and slang expressions tell the reader more about the writer's context, e.g. his social and biographical background.

(178) It's a *pants* film by the way, next time we go 2 the cinema remind me that my boyfriend has bad taste in films! (My emphasis)

Moreover, there are two striking instances of neologisms; one of them is adopted from a TV comedy, while the other one is taken from the children's novel *Harry Potter*. Their use indicates the versatility of SMS and its potential to play and experiment with words and language. The expression *lovely jubbly* in example (179) originates from the British TV comedy *Only Fools and Horses*; it expresses delight, whereas *Mogshead* in example (180) is a town mentioned in *Harry Potter* where the wizard pupils get their sweets.

(179) *Lovely jubbly*! will look forward to seeing you have fun try and make it tonight! (My emphasis)
(180) HI [NAME] ITS MY BDAY PARTY ON SAT.NIGHT 7/7.30 *MOGSHEAD*. BRING FRIENDS IF U LIKE.BUFFET ETC.C U THERE. (My emphasis)

According to Table 3-14 above, there are only 16 instances of taboo expletives in the corpus, this being 0.89 occurrences per 1000 words. Since text messages are even more ephemeral than emails and composed and sent in any state of mind, it could be concluded, according to Kiesler's argumentation, that texting is equally associated with flaming, insensitivity and vulgar behaviour as any other kind of computer-mediated communication (cf. Kiesler 1987; Kiesler et al. 1984 in McCormick and McCormick 1992: 381). This, however, does not apply to this corpus, even though there are a few instances of taboo expletives such as the ones in examples (181) and (182).

(181) HEY? [Name] [name] [name] hav been standin naked on top of that *fuck off* big hill!!! (My emphasis)
(182) *SHIT*! U OK? ANATOMY WAS RANK! WE HAV A LEC AT 2 AND THE EXAM AT 3 IS IN THE LAB I CANT B ARSED WITH IT! (My emphasis)

A second type of flaming, which is not included in the statistics as it does not actually belong to taboo expletives, is what McCormick and McCormick (1992: 390) term "crude flirtation". Quotations (183) and (184) depict examples of crude flirtation. McCormick and McCormick (1992: 390) point out that "those who engag[e] in crude flirtation [are] less concerned about presenting themselves as sensitive, insightful and empathic than the sophisticated users of redefined flirtation and relationship establishment".

(183) so good to see you today, spk to [name]. Thanx for the orgasm! And sex was AMAZING! xxx sleep well

(184) Ill be bak for sum luv wen da union shutz babe cuz l luv u!only kiddin like aye ye ken min?

Like flaming, crude flirtation is the result of writing which is emotionally loaded and text messages with such content share "a social focus and superficial approach to intimacy" (McCormick and McCormick 1992: 390). Yet, whether crude flirtation can actually be classified as flaming or not remains to be questioned. According to Jansen (2002: 172), *to flame* means "to send nasty or insulting messages [...] [which] is usually done in response to someone having broken the rules of netiquette". This definition does not correspond with the above examples (183–184), even though they are of a primitive content. Hence, crude flirtation is a borderline case: it cannot be properly classified as flaming but is, as the term indicates, nevertheless of a certain crudity.

3.5.6 Emoticons as Paralinguistic Modifiers

Emoticons are a characteristic feature of text messages and slightly more common in SMS than in emails, appearing 1.5 times as often. Still, emoticons are not as frequent a feature as Höflich and Rössler (2001: 445) claim, nor is there a great variety of different smileys found. As in the email corpus, the most frequently used emoticon is the original smiley :-). There is also one type of idiosyncratic smiley designed by a computer geek who portrays himself with long hair (*), geek glasses (8) and smiling as can be seen in example (185) as well as in example (186) in which he defines this smiley as "[his] personal one" when asked what it signifies.

(185) Sorry i reply slowly. I've not done much since being back. Jrv catching my breath. Hope to see you online sometime :-) Thanks for your friendship *8-)

(186) Its my own personal one. Its unique. The star is my long hair and the 8 are just geek glasses *8-)

There is even less space and time in text messages than in emails to hedge or modify statements and value them accordingly, which explains why emoticons are more common in this mode of communication than in emails. They substitute paralinguistic cues or can also more clearly

Table 3-15: Emoticons in Text Messages

Emoticons	translation	number	per 1000 words
*8-)	smiley with long hair and geek glasses	4	0.22
:(	frowning/sad smiley	1	0.06
:'(	crying smiley	4	0.22
:-(	frowning/sad smiley	7	0.39
:)	happy smiley	7	0.39
:-)	basic smiley for joking statements / happy smiley	48	2.66
:-/	sceptical smiley	1	0.06
:-D	laughing smiley	3	0.17
:p	smiley sticking its tongue out	1	0.06
:P	smiley sticking its tongue out	1	0.06
:-p	smiley sticking its tongue out	1	0.06
:S	smiley making an incoherent statement	1	0.06
;)	winky smiley for flirtatious/sarcastic remarks	1	0.06
;-)	winky smiley for flirtatious/sarcastic remarks	7	0.39
;o)	winky smiley for flirtatious/sarcastic remarks	1	0.06
total		88	4.88

foreground the hidden meaning behind a simple statement, as this is the case in the above example (185) where *Hope to see you online sometime* was actually motivated by amorous feelings towards the addressee. Thus the :-) smiley not only emphasises the hope to see the person online, but further indicates how much pleasure the sender would derive from that. The fact that the smiley with long hair and geek glasses *8-) was invented and employed by a computer geek further weakens Beutner's (2002: 78) hypothesis that it is usually newbies who make use of smileys (cf. Section 2.5.6, Emoticons as Paralinguistic Modifiers). All in all, it can be concluded that smileys are a characteristic feature of computer-mediated communication even though they are not as frequently used as it is generally believed and only appear 4.88 times per 1000 words as illustrated in Table 3-15.

4 Single Text Messages versus Linked Text Messages

4.1 Definition and Hypothesis

Linked SMS are text messages with more than 160 signs. The earliest mobile phones that offered the feature of linking SMS allowed its user to combine up to three messages. With newer mobile phones as many as 12 messages, i.e. approximately 1800 signs, can be linked. Nevertheless, linked SMS do not seem to be very widespread: interviewees explain that they either do not use linked text messages for economic reasons, or else do not possess a mobile phone with this feature. It is thus especially difficult to collect a corpus of these[1].

The difference between single and linked SMS has not been examined in any study yet. However, the comparison of single and linked text messages promises to be revealing especially in the light of language economy. Many studies claim that language economy in text messages and especially the numerous lexical and syntactic reductions are the result of the restricted number of characters per message (cf. Moraldo 2002: 156). Other studies dispute this argument and maintain that there are other reasons for the highly abbreviated language of text messaging. Siever (2005: 45), for instance, points out that studies from other cultures clearly show that the reduction of signs does not actually result from the restricted space, and he explains that Japanese text messages, for example, provide (if converted) more signs per message and yet the number of signs per SMS does not differ greatly: on average, 50 signs are sent in

1 The corpus of linked SMS that could be collected for this study merely comprises 76 messages. Accordingly, the results are thought to hint towards certain tendencies and must not be regarded as absolute facts. What is more, they cannot be backed up with data from any other studies as the comparison of single and linked SMS in this book constitutes a pilot study. Yet, certain tendencies can still be worked out and provide data for a closer understanding of the nature of text messaging.

Japan and less than 100 in Germany. Likewise, the text messages in this study consist of an average of 96.97 signs (English corpus), 88.11 signs (Swiss-German corpus) and 105.37 signs (German corpus) (cf. Section 1.3, Methodology: Data Collection and Corpora), a result which plainly proves that the maximum number of characters, i.e. 160 characters, is by no means reached. Gerhardinger (2000), Thurlow (2003) and Moraldo (2003) come to similar results, which clearly indicates that the restriction of 160 signs can hardly be the only reason for the syntactically as well as lexically highly reduced language of text messages. Accordingly, Thurlow (2003) explains that "the need for both brevity and speed appears to be motivated less by technological constraints, but rather by discursive demands such as ease of turn-taking and fluidity of social interaction", while Dürscheid (2003: 7) concludes that the creation of reduced sentences already seems to have become a marker of SMS communication. This can be seen where forms of reduction appear even if the writer has still enough signs to use. In his later essays, Moraldo (2004a: 292 and 2004b: 256) also realises that many studies indeed dispute the argument of restricted length being the only factor responsible for language economy in text messages, and he agrees that other factors, such as time pressure and cumbersome typing, may be further reasons for the short texts. Moreover, he asserts that the dialogic structure plays an important role, too: in the different turns, users refer to what has happened before while known things are left out and so, especially in responsive turns, ellipses can be explained by the mutual context which is assumed to be known by both parties.

In order to see how far the length of an SMS indeed influences the language of text messages, it is promising to compare single text messages with linked ones. Interestingly, there is no SMS dialogue in the corpus which includes an embedded linked text message in its thread. Consequently, it can be assumed that linked SMS do not correspond to the demands of text message threads, for it is important to answer quickly and efficiently, i.e. to approximate real-time communication when discussing any given topic via SMS. Linked text messages may, of course, kick off a dialogue, but then shorter replies follow.

Given the different opinions on what factors actually influence the abbreviated language of text messages, the following hypothesis can be put forth: if the restricted length is a major reason for the economy of text messages, one can assume that in linked SMS, there are fewer reductions of any kind. If the linguistics of linked messages is almost the same as that of single messages, however, the proof is given that other factors such as immediacy, speed as well as cumbersome typing[2] are as important, or even more important factors for the highly abbreviated language.

4.2 Salutation and Farewell

As linked SMS do not tend to be part of text message dialogues, it is revealing to see whether they are inclined to contain salutation and farewell formulas more often than single text messages. One would, of course, assume that both formulas are not omitted as frequently as in single text messages, which is indeed the case. As in the email and single SMS corpus, *h*-salutations dominate. Figure 4-1 shows that 35% of the salutations are *h*-salutations whereby the most often used formula is *hey*: 25% of all the messages begin this way. In general, the differences between single and linked text messages are minor. It is, however, remarkable that the *hey*-salutation is significantly more common in linked text messages. What is more, in accordance to the hypothesis above, linked SMS show a lower percentage of messages without salutation formulas: in linked SMS, there are 11% fewer messages without a salutation, i.e. 61% of all the messages lack their salutation formulas, while 72% of the single messages do not contain a greeting.

2 Dürscheid (2003: 7) explains that forms of reduction appear even if the writer is enabled to use the input help T9, which reduces the typing effort of the most often used words. Yet, it is questionable how many users make use of this feature at all, as studies have shown that T9 is not commonly employed (cf. Section 3.2, Historical Background: SMS as Side-Product of Other Mobile Communication Services).

Figure 4-1: Salutations in Single and Linked Text Messages

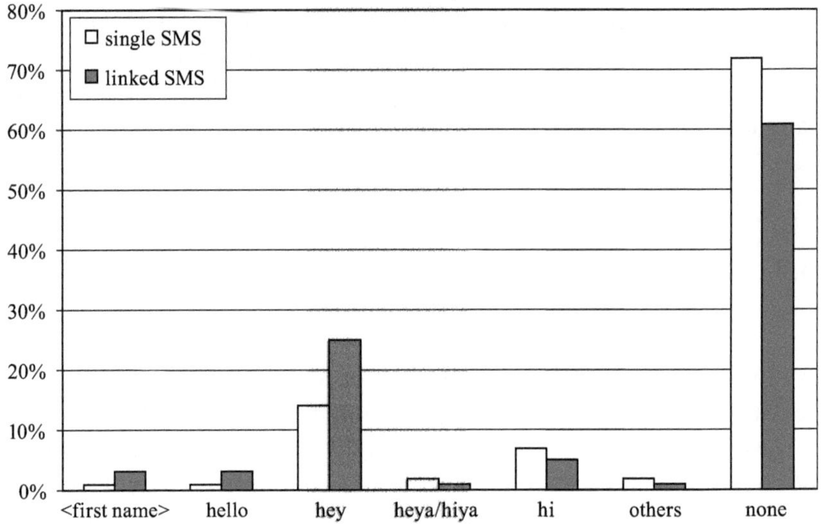

Salutation categories	% in single SMS	% in linked SMS
<first name>	1%	3%
hello	1%	4%
hey	14%	25%
heya/hiya	2%	1%
hi	7%	5%
others	3%	1%
none	72%	61%
total	100%	100%

A similar result can be made out for the leave-taking formulas, as Figure 4-2 indicates. Overall, there are no significant differences. What is noteworthy, however, is that linked SMS are almost twice as often concluded with the name of the sender as well as with the more traditional *love*-formula. Also, there are 16% fewer text messages without a farewell

Figure 4-2: Farewells in Single and Linked Text Messages

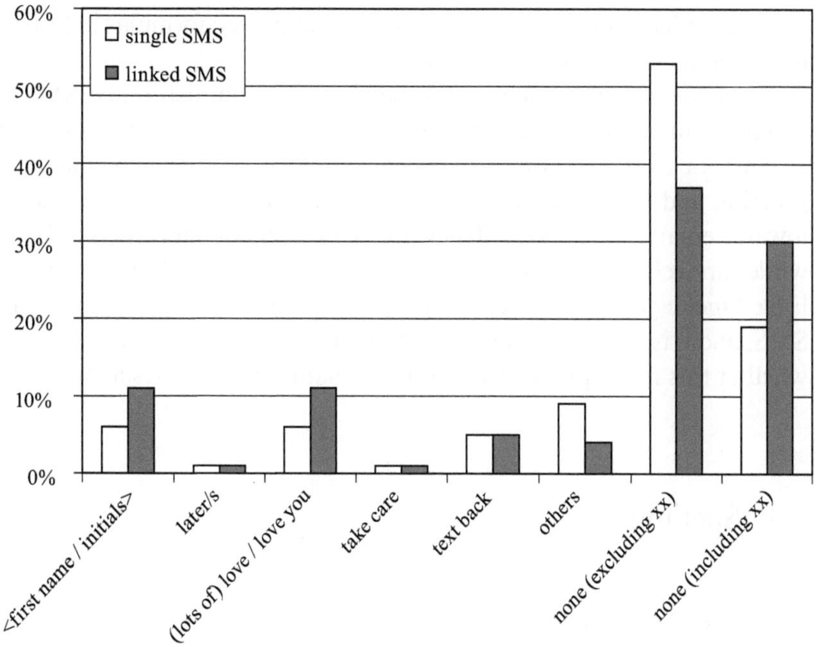

Farewell categories	% in single SMS	% in linked SMS
<first name / initials>	6%	11%
later/s	1%	1%
(lots of) love / love you	6%	11%
take care	1%	1%
text back	5%	5%
others	9%	1%
none (of which xx)	72% (19%)	67% (30%)
total	100%	100%

formula at all, whereas there are 11% more messages which finish off with the kissing symbol *xx*. Hence, it can be concluded that there is a general tendency towards fewer omissions of farewell formulas in linked text messages than in single ones.

These results are in support of the above hypothesis: linked SMS – as far as this corpus is concerned – are not part of any dialogues, or at most kick off a dialogue, and thus contain a smaller percentage of greeting and leave-taking formulas. A careful parallel can be drawn towards emails: here, the tendency not to use salutation formulas is 30%, while farewells are omitted in 15% of all the messages. In this respect, linked messages can be located somewhere between emails and single SMS, and it is certainly interesting to follow this analysis further to see whether this result proves true for the remaining categories as well.

4.3 Spelling

4.3.1 Use of the Upper and Lower Cases

Especially in the light of the fact that linked text messages are not embedded in threads and therefore do not underlie the same time constraint as single text messages, which are more likely to be part of dialogues, one would assume that the percentage of correctly spelt messages in terms of case might be higher in linked SMS than in single ones. The fact that the correlation is exactly the opposite is thus rather surprising: 22% of linked text messages, compared to 35% of single text messages, contain no mistakes in upper and lower case spelling, a ratio of 1 to 1.6. Generally, linked text messages seem to be rather inconsistent in terms of case: the majority of linked SMS, i.e. 68%, use upper and lower case inconsistently – this is almost twice as much as in single text messages. In contrast, there are significantly fewer messages which are all capitalised or exclusively spelt with lower case letters, as Figure 4-3 shows. This result points towards the fact that non-conventional use of case has become a marker of text messaging no matter whether it is a linked message or not.

Figure 4-3: Upper and Lower Case Spelling in Single and Linked Text Messages

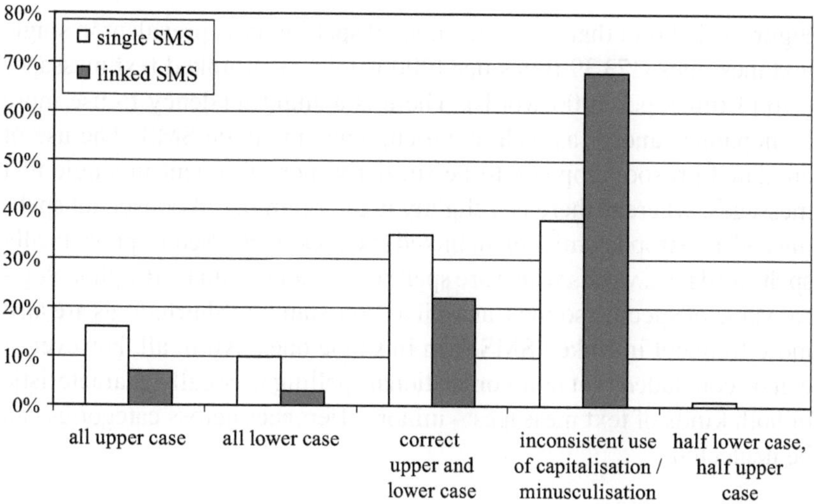

Case	% in single SMS	% in linked SMS
all upper case	16%	7%
all lower case	10%	3%
correct upper and lower case	35%	22%
inconsistent use of capitalisation / minusculisation	38%	68%
half lower case, half upper case	1%	0%
total	100%	100%

4.3.2 Non-Conventional Spelling

Figure 4-4 shows that non-conventional spelling is as prevalent in single text messages (173.39 times per 1000 words) as in linked text messages (176.14 times per 1000 words). There is a slight tendency to use more homophones and *g*- as well as *t/d*-clippings in single SMS. The use of emulated prosody appears to be similarly more frequent in single text messages, whereas there is a slightly greater number of consonant spelling and apostrophe omission in linked text messages. Pseudo-phonetically spelt words, i.e. words which are spelt so as to approximate the phonological value of specific sounds, as well as non-standard shortenings are also more frequent in linked SMS than in single ones. All in all, however, it can be concluded that non-conventional spelling is equally characteristic of both kinds of text messages – minor differences across categories can be neglected.

4.4 Punctuation

Without exception, punctuation marks are considerably less frequent in linked text messages than in single ones despite the fact that there is a significantly greater amount of essential punctuation mark omission in single text messages, as can be seen in Figure 4-5. In fact, punctuation marks are six times more often omitted in single text messages, but at the same time, there are also six times more full stops, nine times more exclamation marks and around ten times more question marks in these messages. Overall, punctuation marks occur almost seven times as often in single text messages (109.58 instances per 1000 words) as in linked ones (16.33 times per 1000 words).

The result of there being fewer punctuation marks in linked text messages than in single ones suggests that the sentences in these messages are generally longer, more of a hypotactic character and commonly more clearly organised so that fewer punctuation marks are needed. Examples (187) to (190) illustrate the use of punctuation marks in single text messages and linked text messages respectively. Comparing these

Figure 4-4: Non-Conventional Spelling in Single and Linked Text Messages

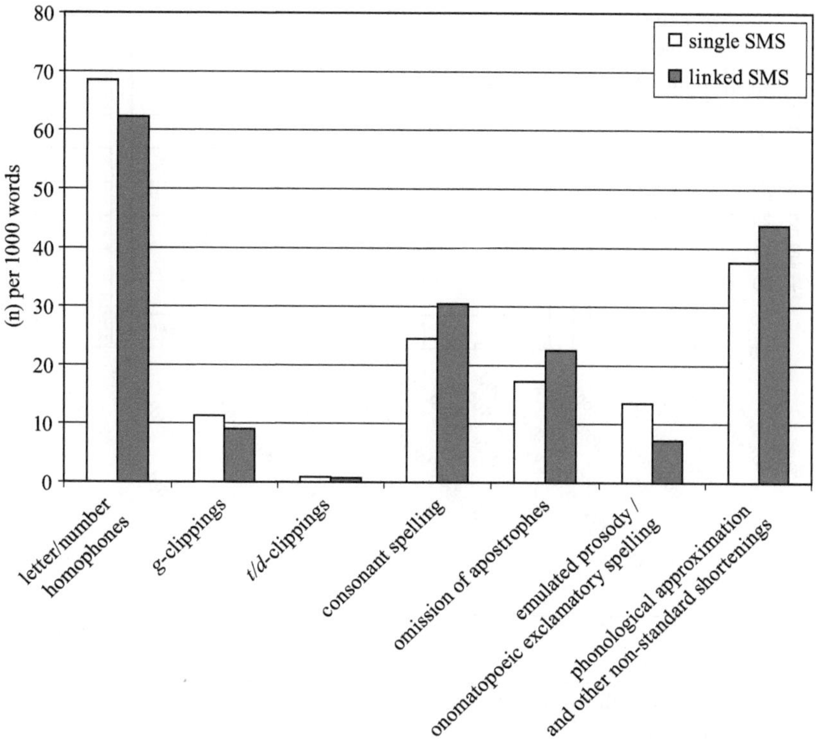

Spelling	per 1000 words in single SMS	per 1000 words in linked SMS
letter/number homophones	68.48	62.28
g-clippings	11.24	9.06
t/d-clippings	0.89	0.70
consonant spelling	24.47	30.44
omission of apostrophes	17.27	22.54
emulated prosody / onomatopoeic exclamatory spelling	13.51	7.20
phonological approximation and other non-standard shortenings	37.53	43.92
total	173.39	176.14

Figure 4-5: Punctuation Marks in Single and Linked Text Messages

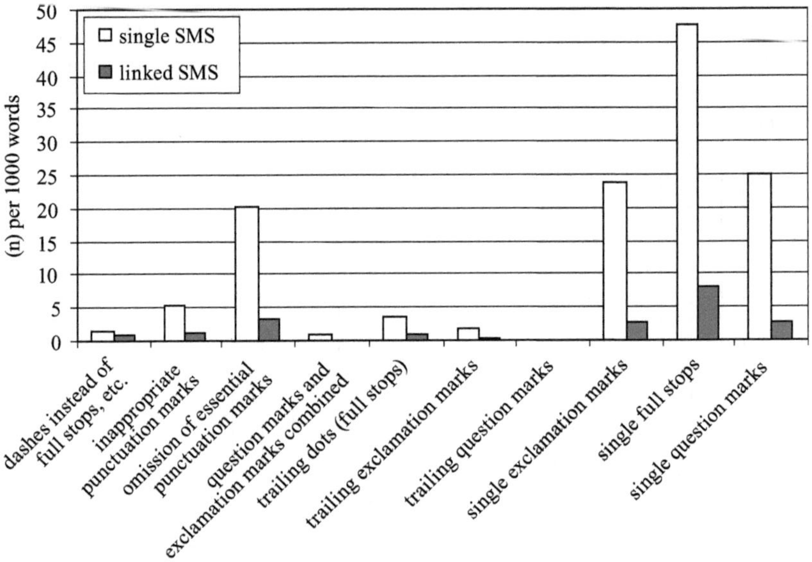

Punctuation marks	per 1000 words in single SMS	per 1000 words in linked SMS
dashes instead of full stops, commas, question or exclamation marks	1.50	0.93
inappropriate punctuation marks	5.43	1.10
omission of essential punctuation marks	20.22	3.31
question marks and exclamation marks combined	0.94	0.04
trailing dots (full stops)	3.60	0.77
trailing exclamation marks	1.66	0.29
trailing question marks	0.11	0.00
single exclamation marks	23.69	2.65
single full stops	47.55	7.94
single question marks	25.10	2.61

messages, it can be clearly made out that the sentences – or sentence fragments – in single text messages are much shorter so that noticeably more punctuation marks are needed. The single text message in example (187) contains 37 words and 7 (fragmented) sentences, the one in example (188) 39 words and 7 sentences; the linked text message in example (189) contains 55 words and 6 sentences, the one in example (190) 87 words and 7 (fragmented) sentences. These examples precisely show that more punctuation marks are necessary in single text messages because their sentence units are smaller, less intricate and more fragmented. In examples (187) and (188), there are 5.43 words per sentence on average, in examples (189) and (190) 10.80 words, i.e. twice as many words.

(187) (1) Sorry was in the shower then got scared by phsyco n jumped out! (2) Shes great as ever. (3) Ill tell her u asked. (4) ye xmas is fine of course. (5) battersea park. (6) Shud b go. (7) tbx x

(188) (1) Wheres dat? (2) Da old sportz thng? (3) I will cum 2dat yea. (4) met me out side it. (5) cya in minz. (6) I aint buyin u shit! (7) U gt cream my hands v dry as a mother tucker!

(189) (1) Hi again. (2) Know nothing about the Nokia 6600 but i know the Nokia camera phones aren't very reliable. (3) Just don't like nokias personally. (4) Look at "google groups" where you can see real people debating and telling their experiences. (5) Compare what you read to a phone u know. (6) I love ericsson and sieman phones.

(190) (1) So, a self-confessed rebel, eh? (2) I knew that the whole "reformed reputable collegeman" image was all just one big act designed to lull the rest of us into a false sense of security while u carry out your evil plots. (3) Well i see through u! (4) Oh, and one more thing; if u absolutely must insist on muscling in on my highly lucrative drug trade once you get to London, then just remember to lay off the East side. (5) Otherwise? (6) There'll be some words. (7) Oh yes.

In view of these results, it might be assumed that emails show an even smaller occurrence of punctuation marks than linked text messages, as a logical conclusion of the analysis would be that sentence structures are again more complex in emails. Surprisingly, however, punctuation marks are five times more frequent in emails than in linked messages. Yet, considering the random example in (191), one can see that emails can even be snatchier in sentence style than single text messages: the email example consists of 83 words and 15 (fragmented) sentences, which corresponds to an average of 5.5 words per sentence.

(191) (1) hi [Name]
(2) lovely to hear from u! (3) that is really nice about [name]. (4) How sweet. (5) have not seen [name]. (6) but i told him about [name]. (7) think is ok. (8) am really getting on with [name] at the moment too. (9) am quite happy. (10) am still not doing enough uni work. (11) i just cant concentrarte or focus. (12) we miss u. (13) some people have been asking bout u so now i have some news to tell them. (14) cool well chat soon and take care.
(15) love [Name]

4.5 Grammatical Features

4.5.1 Syntactic Reductions

Figure 4-6 compares the frequency of elliptic phrases in single and linked text messages. Yet, there are no major differences between the two kinds of messages. All in all, there are 58.85 instances of syntactic reductions per 1000 words in single SMS and almost as many, i.e. 61.82 instances of syntactic reductions per 1000 words, in linked text messages. There are slightly more article/possessive pronoun deletions, auxiliary verb deletions, infinitive *to* marker deletions as well as preposition deletions in the single SMS corpus, while there are more copula deletions, subject pronoun deletions and subject pronoun plus auxiliary verb deletions in the linked SMS corpus. Pronouns other than subject pronouns are equally often omitted in both corpora. This finding is clearly in support of Dürscheid's (2003b: 7) conclusion that the creation of reduced sentences has become a marker of SMS communication regardless of the length of a text message.

In terms of conjunctions, it might be assumed that there are more subordinating sentences in linked text messages, for the hypothesis put forth in Section 4.4, Punctuation, is that there are more hypotactic constructions in linked text messages than in single text messages. There are, however, not only more subordinating clauses, but also more coordinating and asyndetic sentences in linked text messages than in single ones, as can be seen in Figure 4-7.

Figure 4-6: Elliptic Phrases in Single and Linked Text Messages

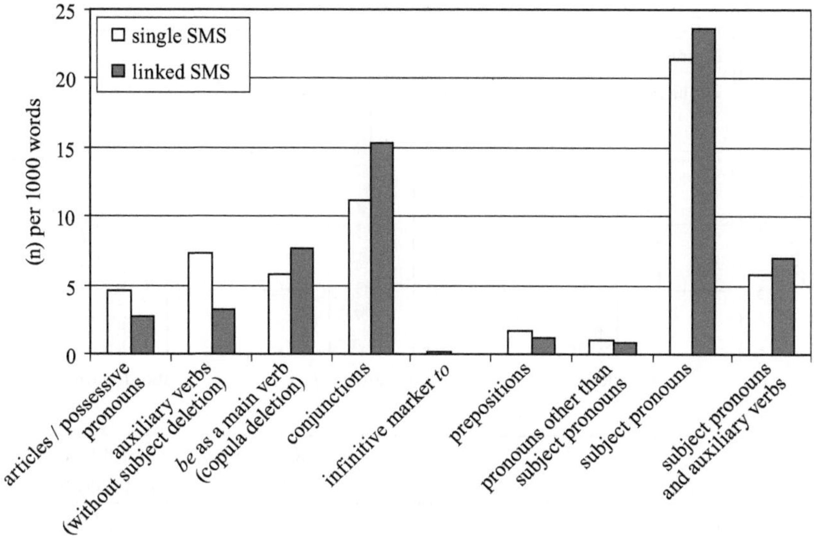

Ellipses	per 1000 words in single SMS	per 1000 words in linked SMS
articles / possessive pronouns	4.65	2.79
auxiliary verbs (without subject deletion)	7.36	3.25
be as a main verb (copula deletion)	5.76	7.67
conjunctions	11.13	15.35
infinitive marker *to*	0.17	0.00
prepositions	1.66	1.16
pronouns other than subject pronouns	0.94	0.93
subject pronouns	21.37	23.70
subject pronouns and auxiliary verbs	5.81	6.97
total	58.85	61.82

Figure 4-7: Conjunctions in Single and Linked Text Messages

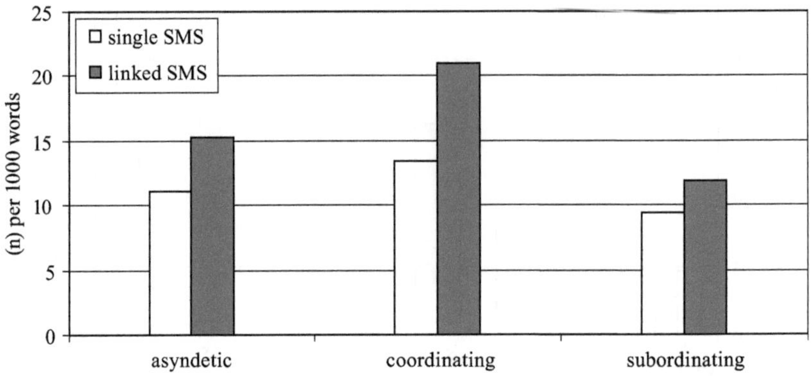

Conjunctions	per 1000 words in single SMS	per 1000 words in linked SMS
asyndetic	11.13	15.34
coordinating	13.45	20.92
subordinating	9.36	12.08

This result again shows that the sentences in linked SMS are longer and more intricate. In single text messages, the sentence structure is not only less complex in terms of subordination, but also in terms of coordination and in respect of asyndetic clauses. In other words, the sentences, as has been shown in examples (187) to (190), are significantly shorter in single text messages and thus none of the conjunctions is as frequently used in single text messages as in linked ones. This finding is also supported by Figure 4-8: all kinds of conjunctions, namely the coordinating *and* and *but* conjunctions as well as the causative and conditional subordinating conjunctions and the remaining coordinating/subordinating conjunctions appear more often in linked text messages. All in all, there are 33.94 conjunctions per 1000 words in the single text message corpus (inclusive asyndetic sentences), whereas there are 48.34 conjunctions per 1000 words in the linked text message corpus (inclusive asyndetic sentences). In other words, there are 1.4 times more conjunctions in linked than in single text messages.

Figure 4-8: The Most Frequent Conjunctions in Single and Linked Text Messages

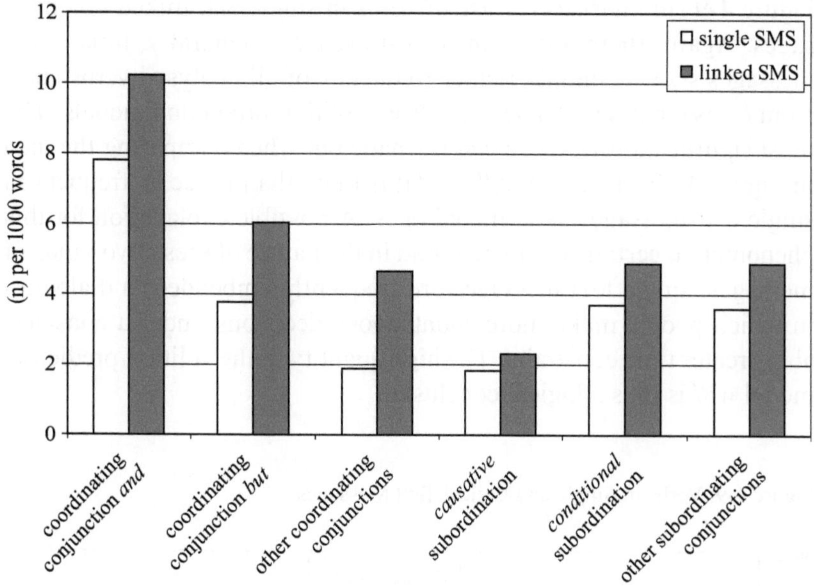

Conjunctions	per 1000 words in single SMS	per 1000 words in linked SMS
coordinating conjunction *and*	7.81	10.23
coordinating conjunction *but*	3.77	6.04
other coordinating conjunctions	1.88	4.65
causative subordination	1.83	2.32
conditional subordination	3.71	4.88
other subordinating conjunctions	3.60	4.88

4.5.2 Verbs

Figure 4-9 summarises the use of verbs in single and linked text messages. Again, there are no major differences. Generally, linked text messages show a slightly higher frequency of all analysed verbs apart from *be* as a main verb and apart from volition/prediction modals. The most significant difference can be made out when comparing the data in Figure 4-10. Here, the *will*-modal is more than twice as frequent in single text messages as in linked ones. A possible explanation for this phenomenon certainly is to be found in the nature of these two kinds of messages: single text messages are frequently embedded in dialogues in which people make more spontaneous decisions under a considerably greater time constraint. The high quantity of the volition/prediction modal *will* is thus a logical conclusion.

Figure 4-9: Verbs in Single and Linked Text Messages

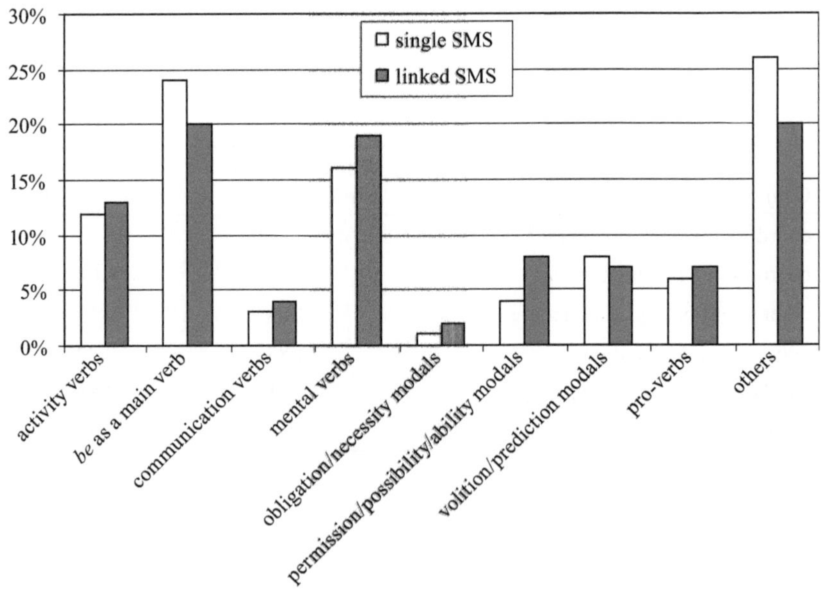

Verbs	% in single SMS	% in linked SMS
activity verbs	12%	13%
be as a main verb	24%	20%
communication verbs	3%	4%
mental verbs	16%	19%
obligation/necessity modals	1%	2%
permission/possibility/ability modals	4%	8%
volition/prediction modals	8%	7%
pro-verbs (e.g. *do, get, put*)	6%	7%
others	26%	20%
total	100%	100%

Figure 4-10: Modal Verbs in Single and Linked Text Messages

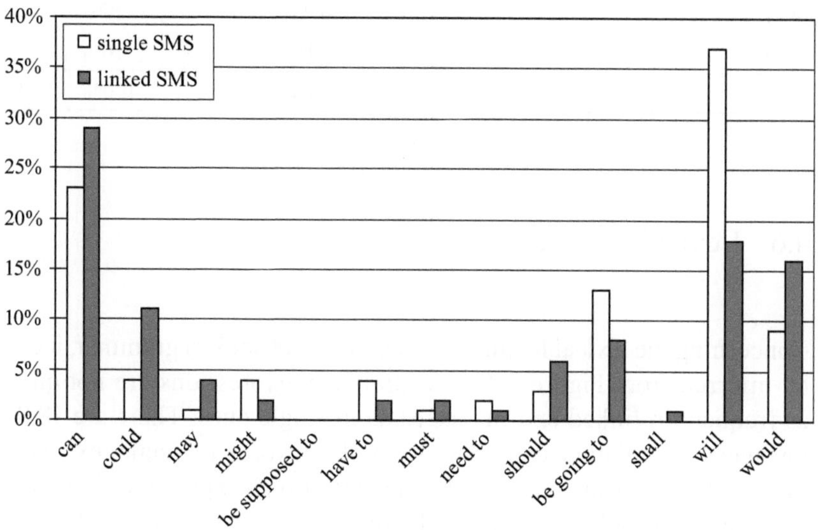

Modals	% in single SMS	% in linked SMS
permission/possibility/ability		
can	23%	29%
could	4%	11%
may	1%	4%
might	4%	2%
obligation/necessity		
be supposed to	0%	0%
have to	4%	2%
must	1%	2%
need to	2%	1%
should	3%	6%
volition/prediction		
be going to	13%	8%
shall	0%	1%
will	37%	18%
would	9%	16%
total	100%	100%

4.6 Lexical Features

Concerning the lexical features characteristic of spoken grammar, it can be inferred from Figure 4-11 that prefatory expressions are not quite as frequent in linked text messages as in single ones. There are 18.84 instances of prefatory expressions per 1000 words in single text messages, while there are only 13.94 instances of such expressions in linked text messages. What is remarkable, however, is the use of the linking adverb *so*, which occurs 1.68 times more often in linked text messages than in single ones, in which the sentences are significantly shorter and often only constitute sentence fragments. This finding supports the hypothesis that linked text messages are more complex in sentence structure and therefore more commonly contain the linking adverb *so*. Quotations (192) and (193) provide examples of the comparably extensive use of the linking adverb *so* in linked text messages.

(192) Hi mate. Sorry not really been in touch. Am on a first aid course at work this week *so* really busy! Will give you a call on Friday to arrange meeting up. Going to a really traditional english pub on Saturday for dinner with my parents *so* that you get to see them too! (My emphasis)

(193) I know about Monday but its Monday today *so* I wanna c u today aint that right *so* where are you now at [names] as long as I get to see you 2day I dont care wheres you go p.s.and da whole of Wednesday we have together,just make sure we see each other doday xxxxxxxx (My emphasis)

Figure 4-11: Prefatory Expressions in Single and Linked Text Messages

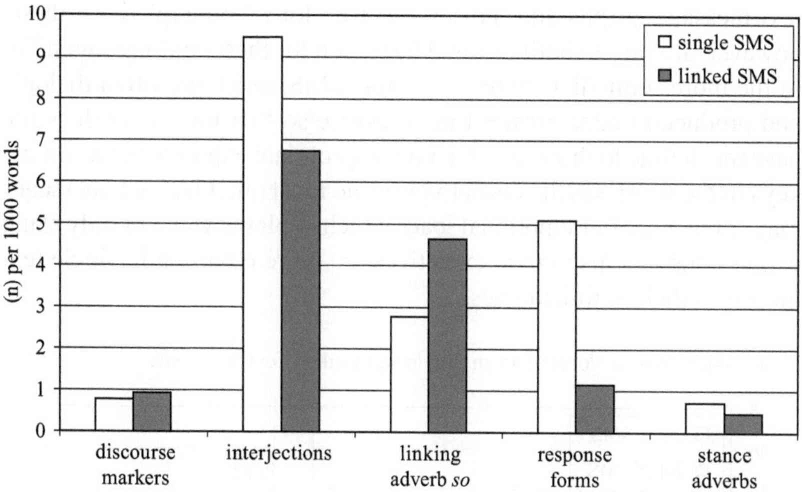

Prefatory expressions	per 1000 words in single SMS	per 1000 words in linked SMS
discourse markers	0.78	0.93
interjections	9.47	6.74
linking adverb *so*	2.77	4.65
response forms	5.10	1.16
stance adverbs	0.72	0.46
total	18.84	13.94

It can further be gathered from Figure 4-11 that interjections and response forms are more common in single text messages than in linked ones. Both interjections as well as response forms are characteristic of dialogues, often being inserted to introduce utterances or to respond to utterances (cf. Biber et al. 1999: 1083). This result again indicates that single text messages are frequently part of dialogues and embedded in threads, which is why these two features are more often used in single SMS than in linked ones, the latter being considerably less dialogic in nature.

Similar to prefatory expressions, spoken vocabulary is more prevalent in single text messages (24.87 instances per 1000 words) than in linked text messages (19.98 instances per 1000 words), as shown in Figure 4-12. The differences, however, are minor and thus not of great significance. There are twice as many slang expressions in the single text message corpus and the same is true for taboo expletives; these, however, are rare in both forms. The reason for these findings again lies in the more immediate nature of single SMS which are often dialogic and produced under greater time pressure so that the writer does not have much time to think about a more appropriate expression, but simply keys in the word which comes to his mind most quickly. Such messages may carry a greater emotional load, which explains why not only slang expressions, but also taboo expletives are more common in single text messages than in linked ones.

Figure 4-12: Spoken Vocabulary in Single and Linked Text Messages

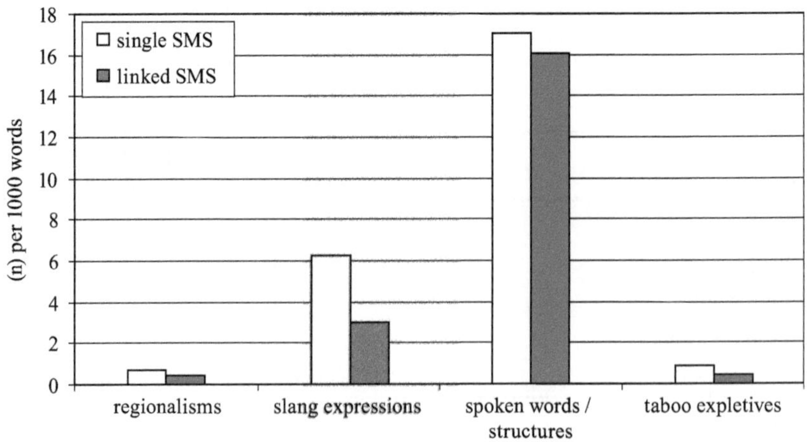

Spoken vocabulary	per 1000 words in single SMS	per 1000 words in linked SMS
regionalisms	0.72	0.46
slang expressions	6.26	3.02
spoken words/structures (excluding prefatory expressions)	17.00	16.04
taboo expletives (excluding moderated expletives)	0.89	0.46
total	24.87	19.98

4.7 Emoticons as Paralinguistic Modifiers

Emoticons are a typical feature of both linked and single text messages. They appear 4.88 times per 1000 words in single text messages and 3.98 times per 1000 words in linked text messages, as Figure 4-13 illustrates. There are no significant differences; the reason for the greater variation of emoticons in single text messages is due to the different size of corpora. If the linked SMS corpus had been as comprehensive as the single one, the variation would certainly have been greater. Hence it can be concluded that emoticons are an equally characteristic feature of single text messages as they are of linked text messages. Again, however, proof is given that, overall, emoticons appear considerably less frequently than one would have assumed, especially when bearing in mind that they are generally associated with computer-mediated communication and text messaging.

Figure 4-13: Emoticons in Single and Linked Text Messages

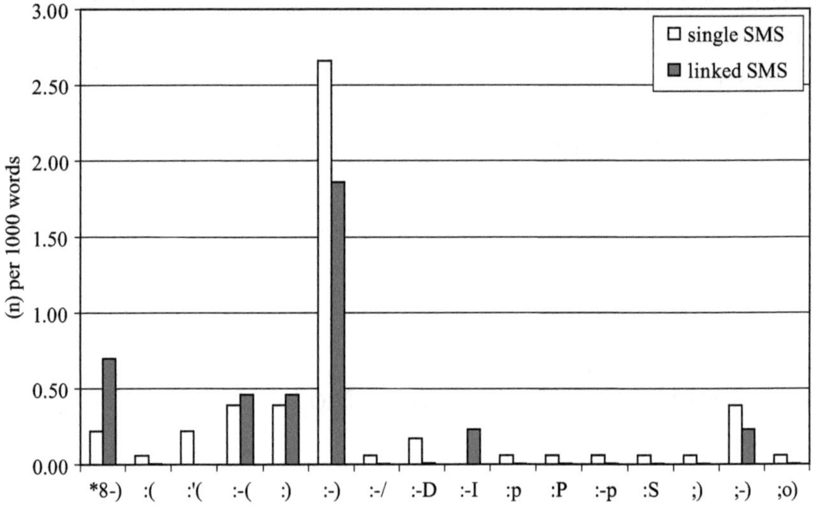

Emoticons	per 1000 words in single SMS	per 1000 words in linked SMS
*8-)	0.22	0.70
:(	0.06	0.00
:'(	0.22	0.00
:-(	0.39	0.46
:)	0.39	0.46
:-)	2.66	1.86
:-/	0.06	0.00
:-D	0.17	0.01
:-I	0.00	0.23
:p	0.06	0.00
:P	0.06	0.00
:-p	0.06	0.00
:S	0.06	0.00
;)	0.06	0.00
;-)	0.39	0.23
;o)	0.06	0.00
total	4.88	3.98

4.8 Summarising Remarks

The analysis and comparison of linked text messages with single ones has brought to light that there are only minor differences between these two kinds of messages. While salutation and farewell formulas are less often omitted in linked SMS, spelling in terms of case is considerably more inconsistent. As far as non-conventional spelling is concerned, however, differences are neglectable. There are significantly less punctuation marks in linked text messages, while syntactic reductions abound in both types of messages. The use of verbs is similar apart from the volition/prediction modal *will*, which is considerably more common in single text messages. As for the lexical features characteristic of conceptual orality, the linking adverb *so* is a recognised element of linked SMS, while interjections as well as response forms are less frequent. Finally, there is no major difference in the use of emoticons. These results clearly show that compared to single text messages, linked text messages

1. are less dialogic,
2. are more complex in sentence structure, and
3. use similar syntactic and lexical reductions in terms of quality as well as in terms of quantity.

In view of that, it can be concluded that the hypothesis proves true that factors others than the limited number of characters play an important role when it comes to explaining the causes of the highly abbreviated language of SMS. Indisputably, the restricted length may initially have led to more extensive reductions on both the lexical as well as the syntactic level. Yet, fostered by the cumbersome typing as well as by the need of brevity and speed and the factor of immediacy, these reductions have in the meantime become a marker of text messaging so that people use them regardless of the available number of characters. Finally, the lexical abbreviations are often also a matter of simplification towards a more accurate representation of speech, approximating the sound system to orthography and thus easing the otherwise rather abstract orthographic conventions (cf. Thurlow 2003). Since text messages score highly on the level of conceptual orality, these abbreviations are a logical conclusion and have little to do with message length.

5 Results

5.1 Major Syntactic and Lexical Features of Emails and Text Messages

Many of the typical features of emails and text messages can be categorised as (material) simplifications of means of expressions (e.g. shortenings, ellipses, parataxis, etc.). Those means of expressions become pragmatically more complex so that their exact function and meaning has to be deduced from the context, which is atypical of written but characteristic of spoken language (Haase et al. 2001: 81). The abundant use of syntactic and lexical reductions in computer-mediated communication has led Ferrara et al. (1991: 21) to conclude that interactive written discourse is a "reduced register" that can be compared to baby talk, foreigner talk and note-taking. This conclusion, however, is too general, for CMC does not constitute a single register; rather, each electronic communication form (e.g. Internet Relay Chat, email, SMS, Instant Messaging, etc.) as well as each type of text (e.g. greetings, goodnight messages, love messages, etc.) is characterised by its own linguistic features. And so, email makes use of such reductions much less than text messages. In fact, almost none of the emails in the corpus do in the slightest resemble baby or foreigner talk[1]. This is different with SMS where

> the challenge of the small screen size and its limited character space [...] as well as the small keypad, has motivated the evolution of an even more abbreviated language than emerged in chat groups and virtual worlds. Some of the same abbreviations appear, either because of their 'obvious' rebus-like potential [...] but the medium has motivated some new forms (Crystal 2001: 229).

1 Even though the language of email messages does not resemble baby or foreigner talk, it may well be that the reduction process is the same (e.g. no articles, etc.). To prove this hypothesis, however, further (contrastive) research would be necessary.

Hence, the language of text messages is characterised by lexical and syntactic reductions, and as Ling and Yttri (2002: 162) correctly state, "there is a certain illicit tinge to SMS messages" which makes it rather difficult for an inexperienced texter to decipher messages such as the following one in example (194), which was composed for UK's nationwide competition to find the fastest mobile texter (Fielding 2002):

(194) IM the Uks fastst,funkyst,speediest,gr8st@txtin!A*!a2nd2nun wndr txter-NO1s qickr thn me!Adctd2txt IM so fst its mgic!! Ha I no ur v.jlus of me!! so:-P
(151 signs)
(meaning: I'm the UK's fastest, funkiest, speediest, greatest at texting! A star! A second to none wonder texter – no one is quicker than me! Addicted to text, I'm so fast, it is magic!! Ha, I know you are very jealous of me!! so nyah!)
(225 signs)

This book, however, reveals that such cryptic messages do not abound: single features like letter and number homophones, clippings, consonant spelling, etc. are employed, but they usually only affect a few words. Still, people who are unfamiliar with text-speak may have difficulty in reading messages such as the one in example (195):

(195) me n mum wer lookin at d holiday brochues last nite n we found a really nice 1 in menorca! Itl b d best hol eva n sumthin 2look 4ward 2 luv ya
(142 signs)
(meaning: Me and mum were looking at the holiday brochures last night and we found a really nice one in Menorca! It'll be the best holiday ever and something to look forward to. Love you)
(176 signs)

The two examples show well that text-speak effectively reduces the length of a message. Both SMS could not have been sent as one message of 160 signs if they had been spelt correctly: the message length is reduced by 33% in example (194) and 16% in example (195). Emoticons also help to reduce message length and are more common in text messages than in emails, where there is enough space to modify or emphasise a statement and indicate the mood that goes with it. But for all that, figures cleary display that emoticons are not as distinctive a feature of emails and text messages as it has often been believed.

Whereas there are around three times as many lexical as syntactic reductions in the text messages (both single and linked ones)[2], the relation is more balanced in emails. Here, syntactic are more frequent than lexical reductions, but only slightly. Overall, both types of reductions are more common in text messages than in emails as Figure 5-1 illustrates. However, "multi-word sentences and response utterances, especially of a stereotyped kind [that] can be reduced to a sequence of initial letters [such as] SWDYT ['So what do you think?'], BCBC ['Beggars can't be choosers'], BTDT ['Been there, done that'], YYSSW ['Yeah, yeah, sure, sure, whatever'], HHOJ ['Ha, ha, only joking']" (Crystal 2001: 229) are neither characteristic of emails nor text messages: even though they are generally associated with computer-mediated communication, they are rarely used. They occur 0.89 times per 1000 words in emails, twice as often in linked SMS (1.59 times) and merely four times as often in single SMS (3.71 times). Most of these multi-word sentences use just the expression *lol*, meaning 'lots of laughter'. Apart from *lol*, there are only a few other multi-word abbreviations such as *afaik* (meaning 'as far as I know') in example (196), *otoh* (meaning 'on the other hand') in example (197) or *tbh* (meaning 'to be honest') in example (198). It can thus be concluded that

> [b]eyond the most obvious impact on linguistic forms of the sociolinguistic maxims, what has been most noticeable about the non-standard items (or 'new' linguistic forms) [...] is how so few of them were especially new or especially incomprehensible [...] In this sense, therefore, claims (both academic and lay) for the impenetrability and exclusivity of SMS language are clearly exaggerated and belie the subtlety and contextuality of discourse (Thurlow 2003).

(196) We're going to [Names] this sat (I don't fancy it but there is this girl I wanna track down and say Hi to) [Name]/[Name] and some others are going *afaik*. (My emphasis)

2 Androutsopoulos and Schmidt (2001) find that their (German) text message corpus contains significantly more syntactic than lexical reductions. This difference between English and German SMS would be worth further research to see how far and why the relation between lexical and syntactic reduction is exactly the opposite in English text messages, and it would also be interesting to cross-linguistically analyse whether there are other points in which they differ.

(197) Aannyy how I work from 1PM-11PM right now so u definatly can't call me then. *OTOH* I am not working all day on Sunday Monday and Tuesday so anytime then is good. (My emphasis)

(198) I have a bit of a fear that he's one of the many gay blokes there causc I heard rumours that it was a gay club and *TBH* the amount of topless hairy scary looking guys that were there last night was quite... uhmm.. scary. (My emphasis)

In terms of orality, there is almost no difference in the frequency of prefatory expressions: they are equally common throughout all corpora. Spoken vocabulary, in contrast, is only half as frequent in emails as in text messages, which indicates that SMS is closer to the conceptually oral language than email. Similar to the spoken language, where no punctuation marks are available, the omission of essential punctuation marks occurs 2.6 times as often in single text messages as in emails. As has been pointed out in Section 4.4, Punctuation, this feature is less prevalent in linked SMS than in single ones, which, however, can be explained by the fact that the sentences in linked text messages are generally longer, more of a hypotactic character and more clearly organised so that fewer punctuation marks are needed and – as a consequence thereof – omitted. The few punctuation marks which are necessary in linked text messages are likely to be put for two reasons: first, in order to guarantee comprehensibility, long and nested sentences tend to be punctuated more than short and simple ones; second, the fewer punctuation marks are needed, the more unlikely it is that they are omitted as the effect of economy would only be minor and thus negligible.

5.2 Spelling Tendencies

As one might expect, the spelling tendencies that have been worked out in Section 3.5.2.3, Non-Conventional Spelling, apply to both single and linked text messages, but can hardly be traced in the email corpus. This finding is well illustrated in Figure 5-2. The differences between single and linked text messages are merely of a minor nature. In linked text messages, there are, for instance, fewer words for which the *night/*

Figure 5-1: Major Syntactic and Lexical Features in Emails and Text Messages

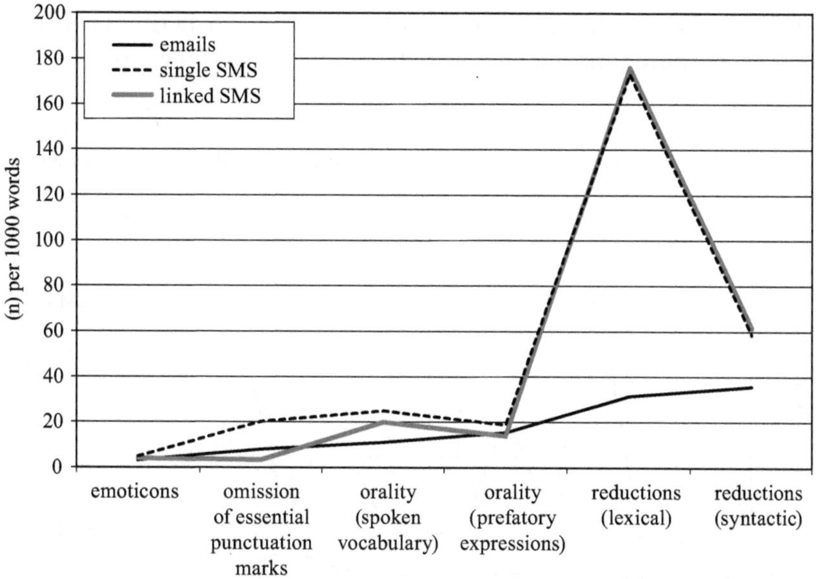

Major syntactic and lexical features	per 1000 words in emails	in single SMS	in linked SMS
emoticons	3.16	4.88	3.98
omission of essential punctuation marks	7.88	20.22	3.31
orality (spoken vocabulary)	11.01	24.87	19.98
orality (prefatory expressions)	15.49	18.84	13.94
reductions (lexical)	31.44	173.39	176.14
reductions (syntactic)	35.78	58.85	61.82

ite-rule applies, but more words which make use of the *double-vowel-single-vowel*-rule. Apart from these most significant distinctions, all rules seem to be equally relevant for both single as well as linked text messages. In the email corpus, however, almost none of these rules appear to be of great importance. In fact, only the *should/u*-rule, followed by the *night/ite*-rule, the *some/u*-rule, the *word-final/z*-rule, the *double-consonant-single-consonant*-rule and the *though/o*-rule show slight tendencies of usage. Yet, compared to the text message corpora, in which these spelling tendencies occur 29.14 and 26.48 times per 1000 words, they are rare and only arise 0.95 times per 1000 words.

Figure 5-2: A Comparison of the Spelling Tendencies in Emails and Text Messages

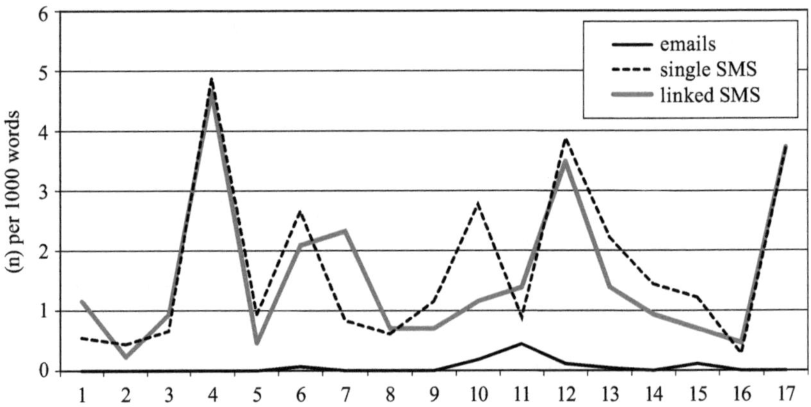

	Spelling rules	per 1000 words in emails	in single SMS	in linked SMS
1	again [e] → /e/	0.00	0.55	1.16
2	any [eni] → /ne/	0.00	0.44	0.23
3	dental fricative [ð] → labiodental fricative [v] → /v/	0.00	0.66	0.93
4	dental fricative [ð] → alveolar plosive [d] → /d/	0.00	4.88	4.65
5	dental fricative [θ] → labiodental fricative [f] → /f/	0.00	0.94	0.46
6	double consonant → single consonant	0.07	2.66	2.09
7	double vowel → single vowel with the corres-ponding pseudo-phonetic value	0.00	0.83	2.32

	Spelling rules	per 1000 words in emails	in single SMS	in linked SMS
8	dropping of word initial /h/	0.00	0.61	0.70
9	intervocalic /r/ → word final /z/	0.00	1.16	0.70
10	night [aɪt] → /ite/	0.18	2.77	1.16
11	should [ʊ] → /u/	0.44	0.89	1.39
12	some [ʌ] → /u/ (+ dropping of word final /e/)	0.11	3.88	3.49
13	though [əʊ] → /o/	0.04	2.22	1.39
14	time [aɪ] → /y/ (+ dropping of word final /e/)	0.00	1.44	0.93
15	word final /s/ → word final /z/	0.11	1.22	0.70
16	word final /y/ → /i/	0.00	0.28	0.46
17	word final [ə] → /a/ / vocalisation of word final /r/ → /a/	0.00	3.71	3.72
	total	0.95	29.14	26.48

The absence of these spelling tendencies in emails is one of the most characteristic differences between emails and text messages. This set of rules can thus be regarded as marker which is unique of text messaging. With a few exceptions, the spelling tendencies that are summarised in Figure 5-2 represent rules according to which words are pseudo-phonetically spelt. The relatively high amount of lexical items which are pseudo-phonetically spelt further indicates the close relation of text messages to conceptual orality. What is more, since non-conventional spelling usually corresponds to these rules, it can be shown that there is a certain recurring pattern when it comes to non-standard and pseudo-phonetic spelling. This proves clearly that the composition of text messages is by far less irregular and random than commonly beheld.

All in all, it can be concluded that the described non-standard written features of emails are even more characteristic of both single and linked text messages. Partly due to the restricted message length, but also because of the small keypad that makes typing cumbersome and time-consuming, users have started to play with language and, more than in emails, there is a marked ludic moment to the composition of SMS. With the newer technology of linked text messages, one might have presumed that a certain counter-tendency back to a less reduced and ludic writing style would be observable. However, these features have already become a marker of SMS and have generally been retained and not dismissed.

5.3 Emoticons as Punctuation Marks

As has been seen in Section 2.5.6, Emoticons as Paralinguistic Modifiers, emoticons are a means to make up for the lack of paralinguistic cues in electronically transmitted messages. Crystal (2002: 131 f.) explains that emoticons are "placed after the final punctuation mark of a sentence". Basically, however, emoticons can be inserted anywhere: at the beginning of a sentence as in example (199), in the middle of a sentence as in example (200), in front or after punctuation marks as in examples (201) and (202), or instead of some form of punctuation as in example (203).

(199) :) HOPE I FIND SOME KEYS :P. I HOPE URE GETTING LOTS OF WORK DONE.ITS OK A BIT COLD HERE.LOVE U LOADS.[NAME] XXXX

(200) I had a really good time too! I had a lovely fryip:) with veggie bacon! I'm soo excited bout stockholm. Wat u doing today? I cdn't be bothered with rowing- so may go home. You going to s'hton? X

(201) ahhhh thanks dude... Whats the odds that I can get Gentoo working with my netgear MA101 USB wireless Lan card ;-)... I bet slim to none...

(202) Do you still do icq or msn BTW? That way it doesn't cost so much to SMS. It's really fun corresponding. I've never really talke like this with anyone. :-)

(203) Waiting at the train station for train into work. The sun is so warm i'm going to miss the next train and just soak what I can up. I forgot how nice it felt ;-)

What has so far been neglected in all studies is the observation that emoticons can substitute punctuation marks, especially full stops at the end of a sentence as long as there are no iterant ones as in example (201) where the impact of the trailing dots is too heavy to be omitted. Indeed, there is a significant tendency to leave out full stops wherever sentence-final emoticons appear. To prove whether this tendency also holds true in a more universal perspective, this feature is further analysed in the German and Swiss German corpora.

Figure 5-3 clearly indicates that full stops are frequently omitted if sentence final emoticons are inserted. Accordingly, 52–76% of all the full stops are replaced by emoticons. In contrast to full stops, exclamation and question marks are hardly substituted which can be explained by the fact that their value is less neutral than that of full stops so that they are not usually left out, but appear in combination with emoticons. Figure 5-4 illustrates this result and also shows that there is just a minority of full stops in combination with emoticons. The only corpus with comparably many emoticon-full-stop combinations is the German email corpus: here, emoticons appear even more frequently together with full stops than in combination with exclamation marks. Nevertheless, the tendency to omit full stops still remains: almost two thirds (62%) of the full stops are omitted whenever sentences end with an emoticon while less than one quarter (23%) of the full stops remain. Similarly, the percentage of emoticons in combination with full stops is relatively high (15%) in German text messages if compared with the percentage of emoticon-exclamation-mark combinations (16%). Yet, there is again a clear tendency towards full stop omission (52%) in case of sentence final emoticons.

Apart from the German email corpus, emoticons most often appear in combination with exclamation marks; they are less frequent in combination with question marks, which is mainly due to the fact that exclamation marks occur more commonly than question marks. Quotation (204) provides a further example of the replacement of a (sentence final) punctuation mark by an emoticon, while examples (205) to (207) illustrate emoticons in combination with punctuation marks.

(204) *bounce* it's not a real kinda "career" job. But I can choose how often I work and if I wanted I could move up the company like my sister did. But its money :-D

(205) As well as that over the last few mouths I've been busy applying for University. I start at Portsmouth Uni on the 03 October 05, to study Law LLB and Criminology. That's the next three years planned. :)

(206) Woo!! I'm going to be good in about 5 min! :-) soooOOOOoooo tired. Wanna chat sometime? Life is fun ATM, how are you finding it?

(207) Oh my!im so brd here!I didn't get 2 sng cuz I culdnt fnd a song!its so borin!lo filipinos!bt no cute ones and my age!wot u doin?eh?seen friends?gud innit?;-)

Overall, the data has plainly shown that emoticons also serve a certain punctuation function: their basic function is to convey their emotional load, but they can also substitute full stops which – in the wake of economy and efficiency – is an important further feature of emoticons. What is more, this function seems to hold true universally: it is not only significant in the English corpora, but also in the German and Swiss-German corpora.

Figure 5-3: Emoticons Substituting Punctuation Marks

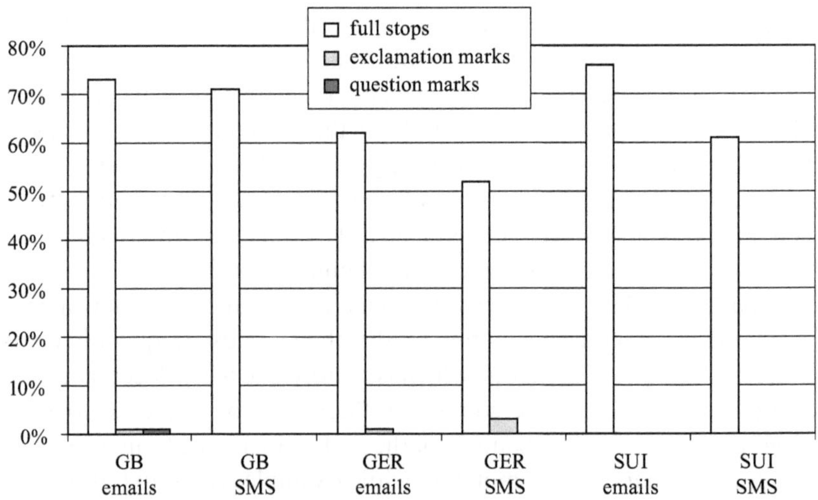

	GB		GER		SUI	
Substitution	emails	SMS	emails	SMS	emails	SMS
full stops	73%	71%	62%	52%	76%	61%
exclamation marks	1%	0%	1%	3%	0%	0%
question marks	1%	0%	0%	0%	0%	0%

Figure 5-4: Emoticons in Addition to Punctuation Marks

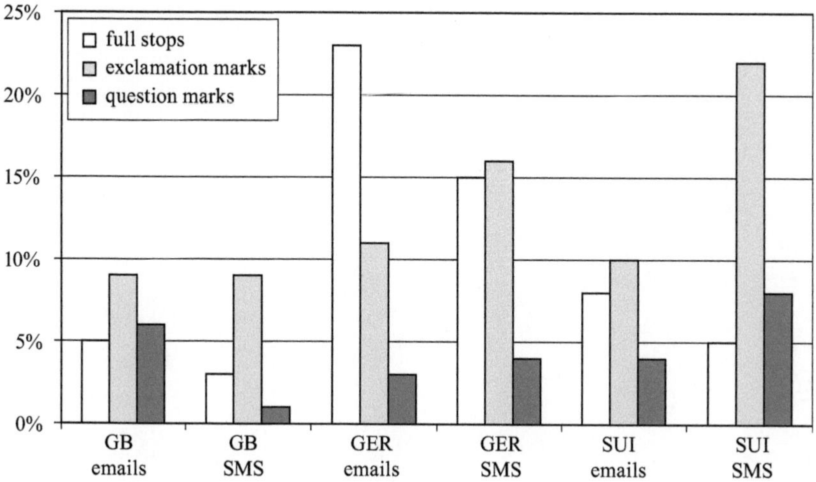

	GB		GER		SUI	
No substitution	emails	SMS	emails	SMS	emails	SMS
full stops	5%	3%	23%	15%	8%	5%
exclamation marks	9%	9%	11%	16%	10%	22%
question marks	6%	1%	3%	4%	4%	8%

6 A Continuing Comparison between Emails and Text Messages

6.1 Email and Text Message Dialogues

Email "has increasingly become recognised as a medium for informal and interactive discourse [...] [and] is often characterised as conversational, although the conditions for a dialogue deviate from those for face-to-face interaction in many respects" (Severinson 1986 quoted from Severinson forthcoming). In contrast to oral interaction, email conversation takes place via a written medium and messages cannot be answered instantly. This time delay between messages is certainly more significant in private than in business interaction; in the latter, people sit in front of the computer all day and are able to answer incoming emails as soon as they receive them. To-and-fro messages can nevertheless be compared to dialogues, each message equalling one turn[1]. However, due to the time-lag, none of the interruptions or overlaps that are typical of spoken dialogues occur, nor is "the clause structure [...] disrupted by false starts, hesitative repetitions or corrections of parts of the sentences, [nor do any] analocutha or otherwise broken and irregular syntactic patterns occur" (Uhlirovà 1994: 278). All the same, emails are "dialogic in character. Although there can be quite a lag before a reply is received, the language in which the reply is often couched is that of a face-to-face conversation" (Crystal 2002: 130), as

1 This equation is not very accurate insofar as messages in asynchronous media usually contain more information than a synchronous turn, i.e. "some email messages effectively convey what would have been communicated through multiple turns in synchronous interaction" (Herring 2001: 620). However, talking about email dialogues (which would be an equally inappropriate term in this respect) leads to the logical consequence of referring to turns when meaning single messages within a thread.

examples (208) to (210) illustrate, which constitute opening lines from three different email dialogues:

(208) Yeah I'm Invoicing today, so should have.
(209) Ohh sorry dude just got this email when I got into the office now.
(210) Hmmm, sounds a bit like gastro-reflux.

There can be minimal dialogues consisting of two turns only, but also extended ones, so-called threads, which contain more than two turns and, in Uhlirovà's (1994: 279) terminology, are made up of "initiatives", "responses" and "follow-ups".

An email message has two features with which cross-message coherence can be created, namely the subject line in the header, which automatically adds the letters *Re* for 'reply', as well as the quoting sign >, which is inserted in front of each line of the previous message. Depending on the number of messages exchanged within the same dialogue, the reply as well as the quoting sign multiply and thus indicate the number of preceding messages. Example (211) depicts a three-turn dialogue, in which the previous two messages are marked with quoting signs so that the whole dialogue can easily be followed.

(211) Email-Dialogue "phone bill?", 23 July 2003[2]

19:05:12	Subject: Re: Re: phone bill? No probs. Just waiting for the bill – not looking forward to that one!
18:34:57	> Subject: Re: phone bill? > hey [name] that is all correct, excepth internet is > £26 and the reason the internet is so expensive is > because it was 12.99 a month since january – I have > been paying by Direct Debit so both you and [Name] > owe me. [Name]
15:46:17	>> Subject: phone bill? >> hey [name], me again. Just a qu. About phone bill: 10.43 line >> rental £25 internet then calls on top of that. Is that correct? >> how come the internet use is so expensive? Hope you are >> well anyway and hi to [name] if you are seeing her soon. >> love [name]

2 These emails are taken from emails others than those included in the analysed corpus.

The mechanism of quoting has advantages for both the receiver and the sender:

> For the receiver, quoting serves to situate the response in a discourse context, e.g. in a situation with multiple ongoing exchanges, and thus facilitates the perception of an extended conversation as coherent. For the sender, it facilitates composition by allowing direct response without having to paraphrase the original message. (Severinson forthcoming)

Moreover, "quoting creates the illusion of adjacency in that it incorporates and juxtaposes (portions of) two turns – an initiation and a response – within a single message" (Herring 2001: 620). Döring (1999: 49) further recognises that quoting approximates monologic writing to a dialogic situation.

There are three different ways of quoting: an old message can follow or precede the reply (indirect quoting), or parts of it can be embedded in the reply (direct quoting)[3]. In example (211), the most recent message is at the top, while the two preceding ones are placed at the bottom, which is probably the most common way of quoting in private correspondence. This hypothesis cannot be proved at this stage, but the fact that none of the emails that could be gathered for this research included embedded or message-preceding quotes leads one to presume that these kinds of quotes are less common than indirect ones that follow the reply. This assumption is also partly backed up by Severinson's (1994: 200) finding that most users "would not include a previous message in a response to a friend or a close colleague". The reason for not embedding quotes or inserting message-preceding ones is mainly due to the fact that email is a hybrid register and so, whereas in writing, quoting is an accepted rhetorical device (even though it may be stylistically disputable), it does not correspond to the norms of spoken interaction, where "the immediate repetition of another party's utterance will often force a sarcastic interpretation" (Severinson 1994: 198). Quoting may further interfere with Grice's Maxims of Quantity, Relevance and Manner and so, when "quoting a lengthy message while addressing only part of it in the response, the contribution will contain more information than

3 The terms indirect/direct quoting are taken over from Beutner (2002: 53).

required (cf. Maxim of Quantity), contain material not relevant to the response (Maxim of Relevance) and be less brief than it should be, or even obscure (Maxim of Manner)" (Severinson forthcoming).

Despite the Gricean conflict, quoting is useful if the message is complex and answered much later, because it helps the reader to contextualise the email by maintaining the reference object. Lee (2003: 317) finds that "e-messages may lack substance in ways that paper texts cannot partly because the subject line makes additional explanation redundant, but largely because the reply function compresses actual time between conversational turns". The greater the time span between two successive contributions, however, the smaller is the resemblance of such written dialogues to spoken ones. Equalling one email message to a turn also implies that such dialogue-messages do not contain greetings or farewells and are marked by elliptic replies. Yet, the more time passes between two messages, the less likely it is that elliptic replies appear and "the more likely [it is that] the response will contain a greeting" (Crystal (2001: 100). Nevertheless, an email message "requires less contextualising content than a letter [...] because the context is already present, as in oral conversation" (Lee 2003: 317). Example (211) shows that the third message does not contain a greeting or a farewell and cross-message coherence is given by the elliptic phrase "No probs" which is an anaphoric answer to the previous message. If the answer had followed a day or two later, the message would probably have been more elaborate, including a greeting and farewell formula as well as a more complete and precise sentence instead of the fragmented and referenceless "No probs". And it might further have contained an enquiry concerning the recipient's well-being as well as an apology for the delayed reply.

Private email dialogues last over a longer period of time and can comprise a couple of hours up to a few days. Theoretically, email can be very close to a synchronous medium if both users, sender and recipient, are online at the same time and answer each other's messages immediately. Practically, however, the time delay in replies proves that email is rather an asynchronous than synchronous mode of communication. This finding also marks a major difference to SMS: text message dialogues take up a few minutes up to a few hours only and SMS are thus significantly closer to the synchronous mode than emails.

In their research, Kasesniemi and Rautiainen (2002: 170–192) have come to the conclusion that in its conversational structure, texting can be compared to online chat insofar as its discourse pattern tends to be dialogic and there can be several messages exchanged within a few minutes. In contrast to email, however, SMS replies lack contextual information, mainly because there is no subject line which gives the reply cross-message coherence, but also because there is no comparable option of quoting. Some mobile phones certainly enable the user to include the previous message into the reply, but this function is hardly used, if at all, for three reasons: first, it would require too much space, second, there are no automatically inserted quoting signs so that it is likely that the recipient would be confused (for he would have difficulty in actually distinguishing between the different messages) and finally, immediacy, which enables the recipient to answer right away, makes quotations unnecessary as the context can be easily remembered.

Two-turn[4] dialogues are the prototype of SMS-communication. They can follow various elementary patterns or adjacency pair sequences such as question–answer, proposal–reaction, request–counter request and serve various communicative purposes like arranging meetings, getting into contact, asking for information, or declaring love (Androutsopoulos and Schmidt 2001: 17). The text messages below (212) provide an example of such a minimal dialogue. They correspond to the prototypical question–answer sequence. What is remarkable, though, is that this dialogue is not concluded but will continue via another medium, i.e. the telephone. It is, in fact, not uncommon to start a communication via an asynchronous medium such as SMS and finish it on the telephone. Calling someone on the mobile phone can be intrusive and so it has become an unwritten law to first contact the person via an asynchronous medium and check whether the addressed person is busy before making a phone call (cf. Freyermuth 2002).

4 To equal one text message with a turn is inaccurate insofar as messages in near-synchronous media may "contain less than a turn, as when for example a sender has more to say than fits in a single message" (Herring 2001: 620) or immediately adds a follow-up. However, talking about SMS dialogues (which would be an equally inappropriate term in this respect) leads to the logical consequence of referring to turns when meaning single messages.

(212) Minimal Dialogue, 28 July 2003

09:41 hey gorgeous.hows the job hunt going? What u up to today, fancy a photo session!? Xxx
09:50 hy you! would love to se u! will call in a bit if u r in and we'll arrange sth. Xxx

Three-turn dialogues usually follow the same pattern as minimal dialogues, but they are expanded by a ritualised phatic-expressive turn such as a confirmation (213) or an expression of gratitude, regret or solace (Androutsopoulos and Schmidt 2001: 17).

(213) Three-Turn Dialogue, 29 July 2003

14:55 hey sweetie.are you in? is it ok to come round in a bit? X
14:59 yep yep, whenever your ready hon xxx
15:03 ok will leave in a bit x

In to-and-fro conversations, people usually build on the content of what has been said before in order to avoid repetition and answer efficiently without losing too much time, for "reaction speed is important" when "messages expand into lengthy dialogues" (Kasesniemi and Rautiainen 2002: 187). The result is elliptic replies; the reply contains new information only while the words of the preceding turn are not repeated. Thus the second turn in example (213) is marked by an adjacent ellipsis which omits the verb phrase 'come round' in *yep yep, [come round] whenever your ready hon.*

In addition to elliptic phrases, there can also be deictic expressions as in example (214) where the pronoun *one* in the first turn marks an anaphoric reference to something that has been communicated before, either via SMS or any other medium or even face-to-face conversation. The actual reference object is not present, but the interlocutors know each other's (situational) context, which allows them to continue their dialogue at the point where they stopped (or interrupted) it before so that they have little difficulty in filling in the semantic gap of such anaphoric references.

(214) Four-Turn Dialogue, 30 July 2003

20:03 Hi, didn't get the one I wanted but they liked me so they have sent my details onto Nhs professionals who will find me one.
20:05 that's good. Don't lose heart...easier said than done but keep pushing-one they know you're serious you'll impress them :-)
20:35 well I think it's better this way cause there doin the searchin for me! Saves me doin it! An ill have a job! How you doin? X
20:49 not so gd. Few problems at home and I've gotta stick around which is unavoidable but was sposed to b going to [name] in 3wks.

Dialogues with more than three turns are of a more complex structure, but their underlying pattern remains the same as that of minimal dialogues. In example (214), for instance, the person informs the addressee about the results concerning the job for which she is looking and later, in her next turn, asks about the addressee's well-being. For each of her turns, she gets a relatively quick reply, first one of solace and then an answer to her question. Similarly, example (215) below is a question-answer dialogue, which is primarily motivated by the mentioned phone bill for which the sender needs the recipient's address, but then ends up in a more phatic conversation, in which the interlocutors enquire after each other's recent activities.

(215) Five-Turn Dialogue, 28 July 2003

19:06 hi [name].how are you?i have the phone bill here.what's your address.I think Ive counted all of mine bu if you disagree,let me know.speak soon x
19:09 Hi. I'm well thank. What have you been up to? My address: [Address].
19:12 Been busy.Learning to drive,working in hotel...still lookin for a jb.how are you getting on?
19:15 Wow quite I then! Hop the driving is coming on. Me, am still looking, had a few rejects and despairing a bit. But going to a wedding at wkend, yay! x
19:20 You'll get there in the end.have fun at the wedding.Good luck for job search. X
19:22 you too. And with driving.Hope [Name] is well too.Take care x

Example (216) illustrates an even more complex dialogue which first asks the addressee whether he wants to come to London for the fireworks, whereupon more small-talk-like turns follow until one of the interlocutors feels too tired and finally concludes the dialogue with good-night wishes.

(216) Eight-Turn Dialogue, 01 November 2005

22:40 Hey [Name] how r u?wanna come to ldn for the fireworks this wknd?How was ur party?
22:45 Well i'm goin2 my dads this weekend, and I cant jus live at ur house!Party was very fun indeed.Hows school?Hows the other luvly ladies?x
22:56 Ahh well some othertime then if youre sure!havent really seen [name] but [name] is good.school was fun. We miss you though!xxx
23:03 If ur not asleep tb cos i'm bored! Def wanna cum down in xmas hols if that's k? How is the legend [name]? Wher r the fireworks? Tb! Xx
23:20 Sorry was in the shower then got scared by phsyco n jumped out! Shes great as ever. Ill tell her u asked. ye xmas is fine of course.battersea park. Shud b go. tbx x
23:23 Aahh gr9 film!Do u not speak 2 [name] much at skool then? I wanna go back on thurs! Wen r u gonna get a job lazy-scar? Do u live most at ur dads or mums?xx
23:25 Yeh I do I just haven't seen her urrrum a bit of both really. Ikno I really need one but im so not that bothered.lol.fallin asleep nite nite [name] xxx
23:34 Haha fair doos then. Nyt Nyt.xx

As in email communication, SMS dialogues with more than two turns are also called threads. Yet, however lengthy these threads are, Androutsopoulos and Schmidt (2001: 15) recognise that a typical SMS dialogue only lasts a few minutes up to one hour at the most. This finding also corresponds to the SMS dialogues in the analysed corpus: 82.5% of them last less than an hour and only 17.5% are of a longer duration, while there is no dialogue which exceeds the period of three hours. The average duration of a dialogue is 34.48 minutes but there is no relation between the duration of a dialogue and the number of turns of which it is made up: none of the dialogues consists of more than eight turns and still, they last between 2 and 169 minutes. In other words, the number of turns is not proportional to the duration of the dialogues, as is shown in Figure 6-1. And so the thread in example (217) is made up of six turns but merely consists of four minutes, while the SMS dialogue in example (218) comprises four turns but takes almost two hours to be completed.

(217) Six-Turn Dialogue, 03 November 2005

14:55 Wher are u?
14:56 At skool. B finished in an hour.
14:57 U wana cum round?
14:57 Yeh wat time?
14:58 Parents are gone from 4 so anytime after tht!
14:59 Kool. Ill cu at bout 4:30. x x x

(218) Four-Turn Dialogue, 03 November 2005

19:30 Hey stranger how are you? [Name] told me you wanted to hear from me, whats up? How's life?
19:58 Lifes gud I spose.Can I ring u 2nyt cos I wanna talk 2 sum1 bout something? Myt as well b u! Hows things in the big city?x x x
21:16 Yes of course you can you I have to ask lol. is everything okay?things are good thanks,cold but goodx
21:29 Yeh.well…no bt we can talk l8r.x x x

Figure 6-1: The Duration of Text Message Dialogues

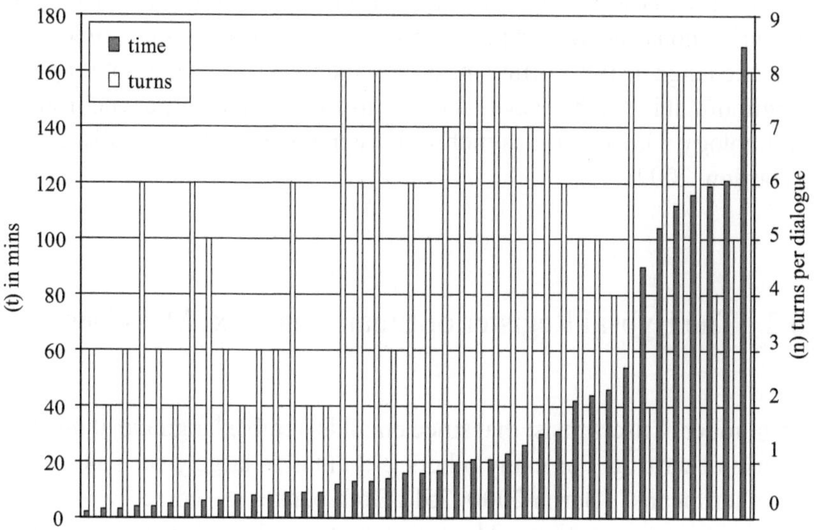

Furthermore, Androutsopoulos and Schmidt (2001: 19) observe that often there are parallel formulations where the recipient copies the structure and elements of the sender. Accordingly, the SMS dialogue in example (215) shows that both dialogue-partners have structured their turns in a similar way: first, they greet each other with a *hi*, then continue to ask about each other's state and talk about their recent activities. Towards the end, they exchange some good wishes and conclude their message with a kiss (symbolised by the letter *x*). Example (213) shows a parallel formulation in terms of semantic content when both partners address each other by nicknames, namely *sweetie* and *hon*, and example (212) contains similar salutation and leave-taking formulas (namely kisses), while the farewell formulas in example (216) are again strikingly identical (*nite nite* versus *Nyt Nyt*). This conformity is a fundamental element that is used to express mutual sentiments; it can help to create a shared frame of ideas while being in different places and situations (Androutsopoulos and Schmidt 2001: 19).

Back-and-forth SMS dialogues that happen within a few minutes demonstrate how an originally asynchronous medium can approximate a synchronous one. If both parties are on reception at the same time, they can exchange near-real-time messages and thus reduce the time-lag to a minimum. It is "in this way that users infuse an ostensibly asynchronous technology with a certain synchronicity in the way they actually use it" (Thurlow 2002).

6.2 The Hybrid Register of Emails and Text Messages

"A number of users insist that since email is durable language, it must be a form of writing. For many others [...] email is largely seen as speech by other means" (Baron 2000: 251). Both parties are correct and incorrect in their assumptions. The register of emails and text messages is hybrid: email and SMS make use of both orality and literacy and "adap[t] the technology of the keyboard [...] to the requirements of talk" (Lee 2003: 323). The missing paralinguistic cues are substituted by various features imitating speech such as emulated prosody (cf. Haase et al. 1997), the employment of capital letters to express loudness or, most obvious, the use of spoken vocabulary as well as the use of emoticons to indicate the mood of a statement. These are features that approximate the written text to conceptual orality by modulating its tone and playing with imitated sounds and emotions. Following Ong (1982), Meise-Kuhn (1998: 231) speaks of secondary orality when it comes to the ludic mode of computer-mediated communication, while Lee (2003: 323) goes even as far as to compare email to reading a musical score because "both sound and emotion reside in the text". Thus, an "experienced" reader, as she would say, can "hear" the frustration in the email in example (219), which expresses disappointment, indicated by the frowning emoticon and capitalisation of the word *frustrated.* In contrast, the reader of the text message in example (220) can hear the excitement of the person in love, which is expressed by the iteration of the letter /o/ in *looove* as well as the repetition of punctuation marks and the smiling emoticon. Similarly, the repeated inflective *slurp* and the grapho-stylistic imitation of a kiss (*mmmwwahh*) in example (221) give the reader an audible idea of the intensity of the kiss, and in example (222), one can hear the loud and relieved sigh of the author, as the inflective *sigh* is not only written in capital letters, but further marked by asterisks, which indicate the performance of the verb.

(219) Due to an error between my mobile phone and my pocket PC (whilst I was trying to back up the numbers in my phone!) I have lost ALL the numbers in my mobile that I've added in the last approx 10 months 8-((. . . . FRUSTRATED
(220) I looove you!!!!! HeHe….. I Never set tired of telling you. Its great :-)
(221) Here's a big snog for you. Slurp slurp mmmwwahh!
(222) My powers of deduction aren't too good so I'm writing to you both in the hope that i've got the right people! *SIGH*

There are, however, also features taken from the written language. Ellipses, for instance, are not always typical of speech and it is mainly situational ellipses (i.e. the deletion of the subject pronoun possibly in combination with an auxiliary verb) which are distinctive of the medially oral language and medially written but conceptually oral texts respectively. Structural ellipses, by contrast, are characteristic of written text types such as diaries, telegrams or mass-media texts (Androutsopoulos and Schmidt 2001: 22; cf. also Dürscheid 2003b: 6). As for the structural ellipses such as article omissions, Dürscheid (2003b: 6) emphasises that these ellipses also occur in other text types, namely in advertisements, information signs or in headlines of newspaper articles. She then wonders what might be new when it comes to article omissions in text messages and concludes that sentences without articles have most often only been read so far, whereas nowadays, the majority of people themselves produce such sentences. Maynor (1994: 53) aptly pinpoints the nature of email and SMS when she explains that "email [the same is true for text messages] obviously employs many features of traditional writing; after all, it is in fact a written medium. Yet it also resembles transcribed speech, with facial gestures added. In addition of approaching the speed of exchange found in conversation, email can be almost as ephemeral as speech" (my comment).

To better understand the linguistics of emails and text messages, it is important to consider the relationship between writing and speech as a continuum and not as a dichotomy because "[c]omputer-mediated communication is not identical to either speech or writing, but selectively and adaptively displays properties of both" (Crystal 2004: 79). "Netspeak", Crystal (2004: 79) argues, "is better seen as written language which has been pulled some way in the direction of speech than as spoken language which has been written down". Unfortunately, Crystal neglects the useful distinction by Koch and Oesterreicher (1985, 1994) between medium and conception (cf. also Schlobinski 2001b and Dürscheid 2004), a distinction which leads to a more profound understanding of the language of emails, text messages and computer-mediated communication in general. The theoretical model by Koch and Oesterreicher, which no longer looks upon orality and literacy as a mere dichotomy, but separates them from the medium or channel so as to be able to regard them as a continuum, is widely applied and discussed within the German language research community (cf. Androutsopoulos and Schmidt 2004; Dürscheid, 2002, 2003, 2004; Moraldo, 2004a). The model can be regarded as a reliable basis on which discourse or text types can be analysed. It was developed in 1985 but is still relatively unknown in the Anglophone research community.

As has been explained in Section 1.6, Emails and Text Messages along the Written/Spoken Continuum: A First Approximation, Koch and Oesterreicher distinguish between two different concepts, namely medial orality and literacy as well as conceptual orality and literacy. The first concept refers to the medium of realisation and is of a dichotomic nature: the realisation of a linguistic element can either be written or spoken. From a medial perspective, an email, for instance, is written while a telephone conversation is spoken. The second concept refers to the ductus and is regarded a continuum. Within this dimension, the boundaries of the written and spoken poles are blurred. In view of that, despite its realisation in the graphic medium, a private letter, for example, is closer to conceptual orality than a text of law. Figure 6-2 illustrates Koch and Oesterreicher's model. It also shows that Koch and Oesterreicher further associate conceptual orality with proximity (i.e. the language of proximity which is conceptually oral) and conceptual literacy with distance (i.e. the language of distance which is conceptually written). This terminology indicates that there are also extra-linguistic factors which influence language choice. Depending on the communicative conditions, "lexical elements can be more or less formal, syntactic structures more or less elaborate, dialectal, sociolectal or colloquial elements can be present to varying degrees and familiarity between sender and recipient always occurs in degrees of more or less proximity" (Weininger and Shield 2004).

Dürscheid (2003: 14–16) criticises the terms proximity/distance for two reasons: first, the terms are problematic as they do not imply any medial associations anymore and second, they do not generally correlate with the corners of the continuum. For instance, a letter that is sent over a wide spatial-temporal distance may still be conceptually oral. She thus recommends not to use the terms proximity/distance and language of proximity/distance respectively, but to use the terms orality and literacy instead. Her criticism is correct, but there is a small detail which should not be neglected. Dürscheid regards proximity and distance only from a spatial-temporal aspect and ignores that there is also an emotional aspect inherent, namely the grade of intimacy or emotional proximity between the partners who communicate with each other. In the light of this broader perspective, the terms proximity and distance seem less problematic even though the medial association is still lacking. Similarly, Androutsopoulos and Schmidt (2004: 56) equate proximity with privacy, intimacy and informality, while Weininger and Shield (2004)

Figure 6-2: Koch and Oesterreicher's Trapezium Model (1994)[5]

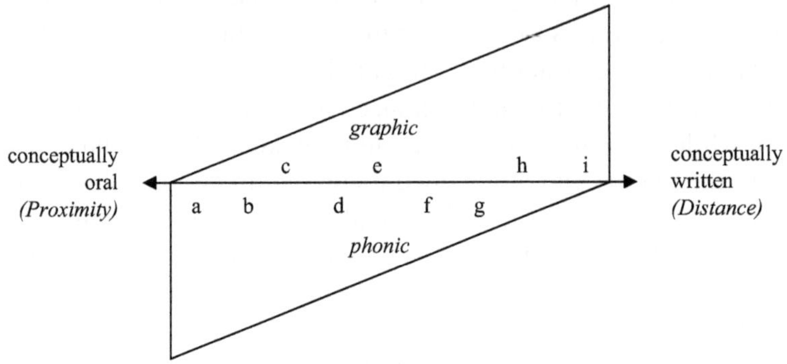

a=private conversation	d=job interview	g=academic lecture
b=phone call	e=published interview	h=editorial
c=private letter	f=sermon	i=text of law

explain that the terms not only condition the geographical or temporal distance between communication partners, but also the "referential, situational, individual, emotional and other types of proximity and distance". Consequently, language is influenced by constitutive factors such as "the degree of familiarity with interlocutor(s), shared knowledge and premises, degree of need to establish common ground, communicative intentions, number of interlocutors and degree of privacy / public accessibility" (Weininger and Shield 2004).

Koch and Oesterreicher (1985) list several factors which constitute the communicative conditions and influence language choice, i.e. the coding strategies, as for instance the choice of lexical features and orality markers as well as the choice of more or less complex syntactic structures. Figure 6-3 shows the same trapezium model complemented by the corresponding factors characteristic of the language of proximity and distance as well as the resulting coding strategies. Adapted to the

5 Koch and Oesterreicher's model is originally in German. To consult the German trapezium model, cf. Koch and Oesterreicher (1994: 588).

Figure 6-3: Communicative Factors and Coding Strategies[6]

Communicative factors

– dialogue	– monologue
– familiar partners	– unfamiliar partners
– face-to-face interaction	– time/space distance
– free choice of subject	– maintain subject
– privacy	– publicity
– unplanned discourse	– planned discourse
– involvement	– detachment
– integrated situation	– assumptive situation
– expressiveness	– technicality
– affectivity	– neutrality
– …	– …

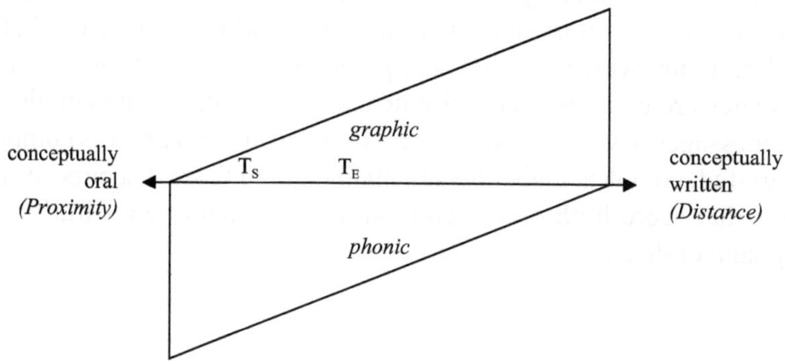

Coding strategies

– process orientated	– object orientated
– provisional	– definitive
less	more
– information density	– information density
– compact	– compact
– integrated	– integrated
– complex	– complex
– elaborate	– elaborate
– planned	– planned
– …	– …

6 For the original model by Koch and Oesterreicher, see Koch and Oesterreicher (1985: 25). Cf. also Koch and Oesterreicher's model adapted by Weininger and Shield (2001) in Weininger and Shield (2004).

corpora discussed in this book, the model also illustrates where private emails (T_E) and text messages (T_S) are to be positioned. They are both closer to conceptual orality than to conceptual literacy, for they include most of the factors constituting the communicative conditions for the language of proximity. According to the corpus analysis above, however, text messages include even more oral features than emails and are thus closer to conceptual orality than emails.

As pointed out, private emails and text messages score high on the communicative factors which constitute the language of proximity. Consequently, their coding strategies are process orientated and provisional. The messages may contain spelling mistakes that would not have been made had they been composed more carefully and reread. In fact, 61%[7] of the informants claim not to reread their messages before sending them, which indicates the great spontaneity with which these messages are composed and dispatched. All in all, private emails and text messages have a smaller information density and are less compact, integrated, complex, elaborate and planned than those discourse or text types that score high on the communicative factors constituting the language of distance.

6.3 Koch and Oesterreicher's Model Adapted to the New Communicative Conditions

Generally, Koch and Oesterreicher's descriptive model provides a good basis to classify discourse and text types. It is even suitable for those electronic messages which did not yet exist at the time the model was advanced. In view of that it does not cause any problems to categorise the text types private email and SMS in the trapezium model of Figure 6-3. Yet, problems arise when trying to gain a deeper understanding of these new types of texts. In various essays, Dürscheid (2002b, 2003,

7 This data is taken from the first survey which was conducted between September 2002 and September 2003. In this survey, 170 students participated.

2004) points out that the model must be extended in order to fit the new communicative conditions. She proposes to distinguish between asynchronous, quasi-synchronous and synchronous communication on the level of medially written communication. An analogous differentiation is made on the level of medially oral communication but here, Dürscheid (2004: 154) distinguishes between synchronous and asynchronous communication only. Medially written and synchronous/quasi-synchronous is, for instance, the online chat, medially written and asynchronous is any communication via email or SMS. Medially oral and synchronous is a telephone conversation, while medially oral but asynchronous is a message on the telephone answering machine. Dürscheid (2004: 154) explains that a central element in this extension of Koch and Oesterreicher's model is the question whether communication proceeds in a synchronous or asynchronous way, for the degree of synchronicity has an influence on the choice of means of expression and this again determines where within the continuum of conceptual orality and literacy a given discourse or text type can be arranged. In other words: the more synchronous the communication, the more it contains features which Koch and Oesterreicher would classify as conceptually oral. Dürscheid's extended model can be seen in Figure 6-4.

Dürscheid (2002b: 58 f.) further proposes to distinguish between electronically (i.e. digitally) and non-electronically (i.e. non-digitally) transmitted linguistic forms on the medially written level. She points out that media linguistics has shown that the transmission path also plays an influence on the linguistic form of a text. An analogous distinction can be made on the medially oral level.

But for all that, problems still arise when classifying electronically transmitted text types. As pointed out in Section 3.5.4.1, Syntactic Reductions, Döring (2002: 14) assumes that the high amount of syntactic ellipses in text messages not only surpasses the rate of word-deletions in other types of written communication, but also in oral communication. In terms of "range of vocabulary", Yates (1996b: 35) argues that "CMC is more akin to writing than speech", for he finds in his research that the lexical density (i.e. "type/token ratio") is highest in writing (50.32), followed by CMC (49.26) and speech (42.29). Thus he concludes that "CMC users package information in text in ways that are more written- than speech-like" (Yates 1996b: 39). However, similar to Döring,

Figure 6-4: Extension of Koch and Oesterreicher's Model by Dürscheid (2004[8])

			conceptually oral ↔ conceptually written
medially oral	synchronous	discourse	
	asynchronous	text	
medially written	synchronous quasi-synchronous	discourse	
	asynchronous	text	

Yates (1996b: 42 f.)[9] also recognises that there are elements which are beyond speech and writing and points out that computer-mediated communication "makes greater use of first and second person pronouns than either speech or writing". The same applies to modals: their use in CMC "is significantly higher than that of either speech or writing, with writing having the lowest usage of all three". Moreover, Maynor (1994: 50) notes that "syntactic features of e-style sometimes reflect informal habits of speech and sometimes are even more simplified than informal speech". Features which are neither characteristic of the oral nor of the written language can also be made out in the German corpora. Bittner (2003: 173) finds that infinitive constructions as the one in example (223) constitute such an element while the copula deletion in example (224) – a feature which is also mentioned by Androutsopoulos and Schmidt (2004: 63) – is characteristic neither of the oral nor of the written language.

8 Dürscheid's model is originally in German. To consult the German model, cf. Dürscheid (2003a: 13 and 2004: 154).

9 Yates (1996: 45 f.) explains the high levels of modals as well as first and second person pronouns by means of Halliday's (1978) theory of the field, tenor and mode and points out that in contrast to other communicative situations, CMC lacks a defined semiotic field, contains a limited semiotic tenor and is complex in terms of the mode.

(223) Drei schaff ich nicht.Vier?*Bitte anklingeln* für bestaetigung.Bussi freu (My emphasis)[10]

(224) Ich nix Kino, ich ø USA! Bis nächste Woche.[11]

These observations shake the assumption that computer-mediated communication can be positioned on the line along the written/spoken continuum as, for instance, Haase et al. (1997), Georgakopoulou (forthcoming), Lee (2003), etc. argue. Rather, it has to be arranged somewhere beyond speech and writing: while making use of both oral and written features, it also has its own features, which are characteristic of neither speech nor writing. The whole concept of the written/spoken continuum must thus be modified and extended by a further level, i.e. a further continuum. Analoguous to Dürscheid's differentiation on the medial level between electronically and non-electronically transmitted linguistic forms, this continuum is constituted by the poles which differentiate between linguistic forms that are conceptually digital and conceptually analogue. According to the new and extended model, which can be seen in Figure 6-5, private emails and text messages can then be arranged in the third quadrant, as they score high (SMS more so than emails) on both the conceptually oral as well as the conceptually digital level.

In his research on Nukulaelae letters, Besnier (1993: 69) observes that "emotion verbs are considerably more frequent in letters than in conversation" and he concludes that "talk about affect is more frequent in letters". This finding might also be true for electronic communication. In fact, both emails and text messages make frequent use of mental verbs. Janney (1996: 202) therefore concludes that there is a

> paradoxical sense of immediacy or nearness sometimes experienced [...] in this medium. There is a remarkable contrast between the cool, disembodied impression made by the e-mail medium itself, and the heated involvement of some of the interaction that it seems to produce. For not fully understood reasons, e-mail seems to encourage displays of affect that would be unusual [...] in normal written correspondence, telephone conversation, or face-to-face interaction.

10 Three I won't manage.Four?Please ring for confirmation.Kiss thrilled

11 I not cinema, I USA! Until next week.

Figure 6-5: Dürscheid's Version of Koch and Oesterreicher's Model Extended

		conceptually oral ⟷ written	conceptually digital ⟷ analogue
medially oral analogue	synchronous		
medially oral digital	synchronous		
	asynchronous		
medially written analogue	asynchronous		
medially written digital	quasi-asynchronous		
	asynchronous	T_S T_E	T_S T_E

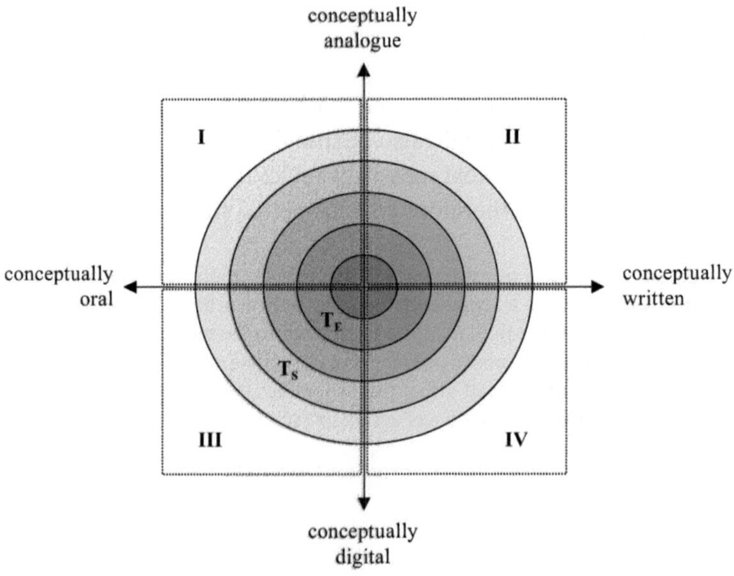

Affection can indeed abound in computer-mediated communication, as the email in example (225) and the text message in example (226) prove (neither of them being written between girl-/boyfriends). This also endows these media a certain touch of superficiality: since affectionate words are used so frequently, they inevitably lose their intimate character. Also, words such as *sweetie, honey, gorgeous, darling, kiss,* or *hugs* are no longer exclusively used between girl-/boyfriends, but expanded to friends and colleagues. Example (226) further shows what an emotional text message can look like if written in the heat of the moment. The sender of this text message has never been experienced as emotional as in an SMS like this one. Even upon calling him immediately, he would never have complained in such a desperate manner or even cried.

(225) hey sweetie. I miss you! sorry about snding texts, I was only told off by [name] yesterday that you have to pay loads to retrieve them. so sorry! anyway be safe and have a wicked time, we love you and can't wait to have you back, love [name] and the world!

(226) I'm a total wreck again :-(can't stop crying. I've never been this sad and hurt before in my life :-(I don't know what to do/think. Life sucks :-(

Yet, as this book does not examine other text types than electronic ones, it can at this stage not be discussed whether mental verbs and affection in general are indeed more frequent in these modes of communication than in any other written or spoken interaction. What can be concluded, however, is that "the sense of the partner's eminent 'interpersonal availability' in e-mail [and text messages] is quite different [...] from feelings of intimacy in other types of human interaction; it has a curiously abstract, virtual quality that is perhaps unique to the email [and SMS] communication process" (Janney 1996: 202, my comments). Furthermore, as has been pointed out in previous studies, "people have a greater tendency to self-disclose on the computer, compared with telephone and face-to-face conversation" (Crystal 2001: 126; cf. also Baron 2000 and Döring 1999).

6.4 Emails and Text Messages Compared to Letter Mail and Telephone Conversation

It has been variously stated that email (the same partly applies for SMS) combines the immediacy of a telephone call with the advantages of a letter: if the recipient is there when he gets the message, he can immediately respond to the incoming message, but is also allowed to take his time to think about what to answer (cf. Graham (1999) and Kinsley (1996)). As with a letter, emails can be stored in files when printed out. Nevertheless, to classify emails and text messages as letters is problematic in two respects: structurally and pragmatically. From a structural perspective, it is inaccurate because the sender or recipient is not necessarily named nor referred to by pronouns so that the narrator/narratee alliance is not always obvious, as this is the case in the email example in quotation (227) and the text message example in quotation (228). Violi (1985: 162), however, points out that these features are crucial elements of a letter: "the presence of the narratee is generally marked by an explicit form, either in the letter's opening ('Dear X'), or in the pronominal structure ('I am writing to tell *you*', etc.)", and she further explains that "an explicitly marked narrator/narratee alliance in the text is a necessary condition for the identification of the genre [letter]" (Violi 1985: 152).

(227) This is quite interesting actually, I've spent qute awhile staring at it now but it doesn't wanna click lol.. Gonna forward it too my folks and ask [Name] here on the voat [= boat] if he can read it too [Name] (My comment)

(228) Snowboard carriage now booked

Pragmatically, email and text messages cannot be categorised as letters, because they are not only composed in situations where one would otherwise have written a letter, but also substitute or complement telephone calls. With regard to immediacy, both media are closer to phone calls than letter writing. Hence, shortly before George W. Bush became President of the US, he explained that for security reasons he would not write private emails any more but use the phone instead (Nolde 2001). Negroponte (1995: 190) finds that "while [email] is not spoken dialogue, it is much closer to speaking than writing" and concludes that this "new quasi-conversational medium is really very different from letter writing". What is more, Katz and Aakhus (2002: 161) postulate the rule that one should first contact the person asynchronously before calling him because asynchronous media are less intrusive than synchronous ones and so, "if the other person responds positively, then one can take the opportunity to respond and perhaps move into more synchronous modes of interaction" (cf. also Freyermuth 2002). Letters, however, can hardly be used to announce a phone call, nor are they always a suitable substitute for emails or text messages because they are not as immediate. Moreover, especially SMS encourages extra-communication, i.e. communication in situations where one would not have communicated before (e.g. pre- or post-face-to-face communication or sending messages out of boredom while waiting for the train). Insofar, neither emails nor text messages can be categorised as letters, nor would it be correct to consider them written phone calls; they do not necessarily substitute phone calls. Emails and text messages are also written to people one would never have contacted or kept in touch with otherwise, and so both of them constitute types of conversation in their own right.

6.5 Gratification and Media Richness Theory

44% of the informants[12] say that they make fewer phone calls since using emails and text messages, while 32% claim to write fewer letters. Only 5% believe that they make more phone calls or write more letters than before, whereas the remaining 51% and 63% indicate no or only little change. This result is very similar to Dimmick et al.'s (2000: 236) finding: comparing the use of email with that of the telephone, they point

12 This data is taken from the first survey which was conducted between September 2002 and September 2003. In this survey, 170 students participated.

out that “nearly one half (48.2%) of the 309 respondents[13] [...] said that they were making fewer long-distance calls, 49.2% said that there had been little or no change, and only 8 respondents said that they had made longer distance calls”. According to Dimmick et al. (2000: 236–238), these “self-reports clearly indicate a displacement effect”. They further argue that “there is a moderately strong competition between email and the telephone”: while “the telephone is superior for the sociability gratifications that are highly affective uses in personal relationships, such as expressing emotions and affections, giving advice, exchanging information, and providing companionship”, email exhibits the following advantages: it “fits peoples’ work schedules”, simplifies “communication with people in different time zones” and helps to “kee[p] in contact with people who live far away” or those who “you don’t have time to see in person” (Dimmick et al. 2000: 240).

An important factor for the choice of media is the criterion of media richness. According to Daft and Lengel (1984, 1986), there are four factors which determine media richness, these being

1. the ability of the medium to transmit multiple cues,
2. the immediacy of feedback,
3. the use of natural language,
4. the personal focus of the medium.

(Daft and Lengel 1984, 1986 in Spears et al. 2001: 605)

This means that “richer media have more cues [...] instant feedback [...] are able to cope with a variety of language and hence are better at conveying meaning [and] allow personal feelings and emotions to be conveyed” (Rudy 1996: 200). With regard to these four criteria, email as well as SMS can be considered leaner media than any spoken interaction mainly because they lack the paralinguistic cues. These cues cannot be fully substituted by graphic icons, i.e. emoticons, or emulated prosody and as such are “best suited to fairly routine and low-level

13 The majority of Dimmick et al.’s (2000: 236) informants (98.3%) were between 18 and 30 years old and all of them were educated: “31% had a bachelor’s degree, 13.9% had completed some graduate work, and 18.1% had completed graduate school”.

communication uses" (Spears et al. 2001: 605). This also explains why it is widely felt that delicate and negative content, such as the breaking off of a relationship as well as any other kind of bad news, should not be communicated via these media, but are best told in oral interaction: "the voice channel carries not only the signal but all the attendant features that make it have the traits of being understanding, deliberate, compassionate, or forgiving" (Negroponte 1995: 147). In other words, "embedded in sound is information about feelings" (Negroponte 1995: 148). Moreover, the possibility of immediate reaction is especially important when emotional support is needed. Longmate and Barber (2002: 72) further point out that "rich media will be chosen in situations where the sender wishes to reduce equivocality i.e. provide a careful explanation for an unclear situation. Leaner media such as written communication will be chosen in situations where the sender wishes to reduce uncertainty by providing large amounts of detailed information".

The gratification and media richness theory explains why the telephone has not been ousted by email. A new medium only has the potential of replacing an older one if it "gratifies the same needs as the older medium" and if "the newer form [is] superior to the older form" (Dimmick et al. 2000: 234). Email, however, does not meet any of the demands and so, "even among its heaviest users, electronic mail has not become a substitute for […] telephone calls" (Rice and Case 1983 quoted from McCormick and McCormick 1992: 400). A certain displacement has nevertheless taken place, namely for messages of a more trivial nature or minor importance. But it is dubitable whether all this information would actually have been communicated at all before the advent of email and in this respect, the displacement must be relativised.

SMS is similar to email, but it is more immediate and "a highly interactive medium" (Longmate and Barber 2002: 78), for it is a mobile medium and the access to mobile phones is generally easier than to computers. Hence, Leung and Wei (2000: 308) find that "the use of cellular phones on buses, cars, and trains or in malls and restaurants is strongly linked to mobility and immediate access gratifications". They establish seven gratification factors, these being fashion/status, affection/sociability, relaxation, mobility, immediate access, instrumentality and reassurance, concluding that "mobility, immediacy, and instrumentality are found the strongest instrumental motives in predicting the use of cellular phones". Whereas Baron (2000: 188) argues that "the profusion of cellular phones [...] is undercutting email", Höflich and Rössler (2001: 456) do not see any connection between SMS and email and state that SMS is neither considered a substitute nor does it indicate – as 'email for the mobile phone' – any general preference for this type of communication. Similar to Leung and Wei (2000), they define five gratification factors for SMS, namely reassurance, contact, immediacy, help and fun, and further emphasise the ludic moment that is inherent in this medium (Höflich and Rössler 2001: 455).

Dimmick et al. (2000: 240) explain that "[a] new medium survives and prospers by providing utility or gratification to its patrons. In doing so, it may have effects on existing media by providing new solutions to old needs or to more contemporary demands. One way of defining a medium's niche is as that region of the resource space where it outcompetes similar media". With respect to this statement, however, Höflich and Rössler (2001: 458) recognise that it is too early to actually tell which status SMS can be allocated in everyday communication and how far it will influence the use of older media. This is certainly true because SMS technology has just started to get off the ground and its possibilities will not be exhausted in the near future. Hence, it remains to be seen in which direction the use of SMS will lead; whether there will be a serious competition between SMS and email and how far it will change the use of telephone calls. All the same, Ling and Yttri (2002) have worked out four major gratifications of SMS over email, namely "high transportability (more so than email), reasonable affordability (more so than email), good adaptability (e.g. also voice-phone) [...] and general suitability (e.g. it is quiet, discrete)" (Ling and Yttri 2002 quoted from Thurlow 2002).

As for the advantages of SMS over telephone calls, Grinter and Eldridge (2001: 231) find that teenagers prefer text messaging for two main reasons:

> First, they can avoid the time, and cost, of all the various conversational protocols required before they can ask a question or get to the reason why they called. The character limit of the messages themselves makes this terse and otherwise rude behaviour completely acceptable. Second, they can avoid the other person going 'off topic' and making the conversation even longer than planned. Again, the character limit forces both sender and respondent to stick to the topic. In summary, then, text

> messaging is quick because the teenagers know the interface and the terseness of the medium speeds up the exchange and focuses it.

Grinter and Eldridge (2001: 231–233) conclude that teenagers favour text messaging over other media because texting is cheaper, easier and more convenient. They find it a good alternative to contact someone who finds telephone conversation difficult, or else use text messages "to avoid making conversation with people they do not know well". Also, with text messaging, silent communication is possible that does not disturb other people. Some physical locations such as cinemas, libraries or museums require discretion. From such places, it would be unacceptable to make phone calls. Texting, in contrast, "provides anytime, anywhere contact with other people, but in public settings, it is a way to be discrete".

An important advantage of email and SMS over any spoken interaction is their asynchrony. Ling and Yttri (2002: 159) explain that "on an interpersonal level, SMS is an important medium for filtering communications; it also allows one to reflect while composing a message. In Goffman's terms, the indirect nature of text messages [and emails] allows one to arrange 'face', that is, it allows one to consider the effect that is desired in the message" (my comment). In this context, McCormick and McCormick (1992: 394) add that "because people cannot be judged by their looks or other superficialities when they interact on the computer, electronic flirtation and overtures of friendship may feel safer than face-to-face interaction".

The advantages of email over older media were recognised early: already in 1978, Joseph Licklider and Albert Vezza, two employees of ARPANET, pointed out the advantages of email over the letter and the telephone. Their conclusion is still relevant today and also applies to SMS when they explain that

> one could write tersely and type imperfectly, even to an older person in a superior position and even to a person one did not know very well, and the recipient took no offence [...] Among the advantages of the network message services over the telephone were the fact that one could proceed immediately to the point without having to engage in small talk first, that the message services produced a preservable record, and that the sender and receiver did not have to be available at the same time (Licklider and Vezza 1978: 44).

There are various different factors which influence media choice and Decker recognises that “to write rather than call, to call rather than write, to write or not by e-mail [or SMS] are decisions made with respect to the time frame in which the message must be conveyed, the status of interpersonal relations, rhetorical advantages and disadvantages, and many other considerations more or less complicated” (Decker 1998 quoted from Höflich 2003: 58, my comment).

7 Text Messages as the Renaissance of Telegrams

Text messages have often been compared to telegrams. Schwitalla (2002: 33) speaks of a comeback of the telegram and points out that media which have been outstripped by newer technical developments may have a comeback as, for instance, the telegram, a text not on paper but on the mobile phone. In an interview, Höflich also considers the advent of SMS as a renaissance of telegrams (Stroh 2004). And Schlobinski et al. (2001: 22) similarly compare text messages to this earliest mode of electrical communication in writing when it comes to the linguistic shape of SMS, while text messages may even be directly referred to telegrams as the following prefabricated German SMS saying illustrates (229):

(229) Da ich nicht singen kann, schick ich dir jetzt einfach ein SMS-Grußtelegramm: Happy Birthday to you! (Anonymous 2006)[1]

The following sections provide an insight into the nature of telegrams and examine how far text messages indeed resemble telegrams and in what ways they are different. The telegram corpus consists of messages which were all composed before 1950. They were written among UK citizens. There are only private telegrams included but no wedding greetings[2]. The analysis of telegrams and SMS promises to be revealing, yet the results must be handled with care as the target group is different: while the majority of the users of text messages are young people, the principal users of telegrams were older (cf. Siever 2005: 144). Accordingly, the messages in the SMS corpus were exclusively composed by students aged between 17 and 27, while the telegrams were written by

1 As I cannot sing, I simply send you an SMS greeting telegram: Happy Birthday to you!

2 The reason for not including wedding greeting telegrams is that wedding greetings abound. The aim, however, was to collect a heterogeneous corpus of various different text types, all of them being of private nature still.

older people. Nevertheless, certain tendencies can still be worked out, and the results will show how far the equalisation of text messages with telegrams is justified.

7.1 Definition

Albert Einstein once defined a telegraph "a kind of a very, very long cat. You pull his tail in New York and his head is meowing in Los Angeles". In more serious terms, a telegraph is an electrical[3] communication system over which information can be transmitted in the form of Morse code. The message sent by telegraph is called a telegram. Telegrams usually consist of rather short messages: since the cost depended upon the number of words, a unique writing style, the so-called telegraphese, developed. People tried to limit the wording so as to keep the messages as cheap as possible and frequently omitted those words which could be easily understood from the context.

A telegram passed several people until it reached its addressee. Classically, the sender filled in a form on which he wrote his message, which was then given to a telegraphist who manually sent the message over the Morse-key. At the end of the wire, another telegraphist decoded the message and wrote it on a telegram form, which was finally given to the addressee. With the Hughes printer, invented by David Edward Hughes in 1855, the letters could directly be sent over key pads and were then automatically printed on stripes of paper which only needed to be cut and pasted on a form. These telegrams often had an unusual typeface: no punctuation marks, exclusive upper or lower case spelling[4],

3 There is, of course, also the mechanical telegraph which existed prior to the electrical one. Accordingly "the first communications system able to transmit any information was built by Claude Chappe [...] It covered the distance of 210 km between Paris and Lille. The first telegram over this line was sent on 15 August 1794" (Solymar 1999: 22). The focus of this section, however, is on the electrical telegraph, which marked the beginning of electrical communication and is thus more related to today's electronical communication technology than its mechanical pendant.

or transmission errors which resulted in spelling mistakes. The fact that several people read the message until it reached its addressee was unfavourable for intimate conversations and so, phatic communication remained restricted to greetings and congratulations (Schwitalla 2002: 34–46).

Apart from the technical quantum jump, the advantages of text messages over telegrams are mainly cost and convenience. Text messages are less laborious to dispatch than telegrams. They are no longer stationary and can be sent and received anywhere, provided reception is possible. What is more, all this can be done without any mediators. Schwitalla (2002: 54) regards the absence of mediators as one of the central differences: the privacy of SMS, he explains, produces the major functional and stylistic differences between the two media. He therefore considers text messages a medial benefit for privacy.

7.2 Historical Background

"The first proposal to use electricity for sending messages [...] was made in an anonymous letter to Scotts Magazine in February 1753. The author signed himself as C.M." (Solymar 1999: 50 f.). Yet, almost another century was to elapse until a first commercial telegraph was established: it was a telegraph cable "along the railway line from Paddington to West Drayton, a distance of thirteen and a half miles", which opened in 1839 (Solymar 1999: 53). The telegraph was designed by the British physicist Sir Charles Wheatstone in cooperation with the British engineer Sir William F. Cooke. "The man instrumental in setting up the electric telegraph was Samuel Morse, Professor of the Literature of Art and Design at the University College of New York at the time [...] He has been credited with the invention of the Morse code which soon spread all over the

4 In the telegram corpus analysed for this study, 67% of the messages are wholly capitalised whereas none of the messages is spelt in all lower case letters. 82% of the messages contain no punctuation.

world" (Solymar 1999: 54). By means of this code, the messages could be transmitted in forms of electrical impulses. The Morse code

> was not far from an *optimum* code in the sense that it could transmit a given text in minimum time. [Morse] achieved that by the simple method of analysing a certain text and attaching shorter codes to letters occurring more often and longer codes for those coming up less frequently. The single dot, for example, was reserved for *e*, the letter occurring with the greatest frequency (Solymar 1999: 54).

In 1845, the first telegraph service, the Electric Telegraph Company, was registered in the United Kingdom, and in 1859, the London District Telegraph Company aimed at wiring up London over the roofs (cf. Solymar 1999: 58). Before the invention of the telephone in 1876 by Alexander Graham Bell and its general access by the public, telegraphy was the quickest method to transmit messages – it was faster than the postal service. Despite the costs, it quickly became popular so that the messages that were dispatched in Great Britain increased from only around one million in 1858 to over 25 million in 1878 (cf. Solymar 1999: 58–66).

In Britain, the peak was reached in 1913 when a total of 82 million telegrams were transmitted. From then on, telegrams lost status and their usage declined regularly until 1939, when only 50 million telegrams were sent in the United Kingdom. During the Second World War, the telegram service was suspended. It was revived afterwards, but the number of the telegrams has been shrinking ever since. In 1981, telegram service became part of British Telecom and was finally taken over by TelegramsOnline in 2003 who – in the age of new media where transmission has become much faster – proudly reported that they had sent over two million UK telegrams in 2005 (TelegramsOnline 2006).

7.3 Linguistic Characteristics of Telegrams in Comparison with Text Messages

7.3.1 Length

The average length of text messages and telegrams is surprisingly similar – both in terms of words as well as in terms of characters. While text messages are restricted to 160 signs[5], the restriction in telegrams was cost. For this reason, both types of messages are relatively short, as Figure 7-1 shows. While the mean length of telegrams is 14.6 words or, in terms of characters, 78.8 characters, text messages are slightly longer with an average of 19.9 words, i.e. 96.97 characters.

This result is similar to Veredarius' finding: he concludes that German telegrams have an average length of 14.2 words and explains that after the price-categories of telegrams changed from a distance-tax to a word-tax, the number of words dropped from 18.3 to 14.2 words (Veredarius 1885 in Schwitalla 2002: 37). Veredarius' count is also on a par with that of my German telegram corpus, which comprises 120 messages: here, the mean length of telegrams is 11.5 words, i.e. 72.4 characters. According to these results, telegrams and text messages seem to be very similar in length. Even in terms of the minimum and maximum numbers of words, text messages are very much like telegrams. While the shortest SMS consists of one word (i.e. one emoticon) only, the shortest telegram consists of two words (230); the longest SMS, in contrast, comprises 40 words, while the longest telegram counts 52 words (231).

(230) Successful [NAME]
(231) SHALL NOT BE ABLE TO GO TO PARIS TOMORROW PLEASE CONSIDER WHETHER IT WOULD BE ADVANTAGE TO MAKE THIS EXCUSE FOR

5 For reasons of consistency, the comparison includes single text messages only. The differences between single and linked text messages have been analysed in Chapter 4, Single Text Messages versus Linked Text Messages. A comparison between linked text messages and telegrams is futile, for the differences between single text messages and linked ones are too minor as to significantly influence the results in the following sections.

Figure 7-1: The Length of Text Messages and Telegrams

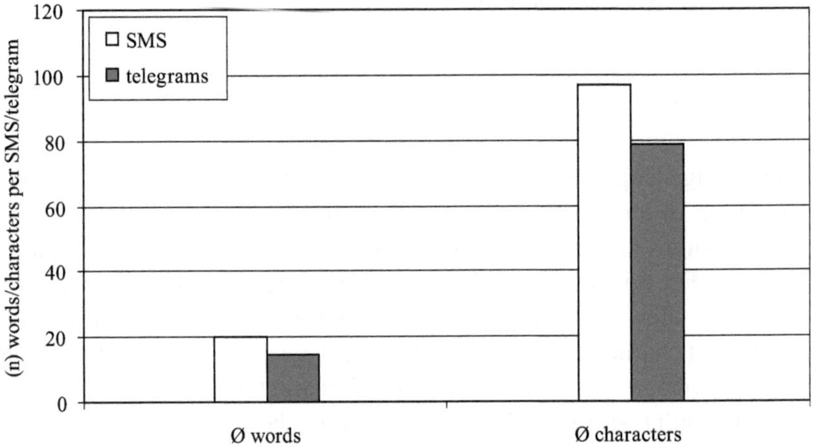

Length	Ø words	Ø characters
SMS	19.90	96.97
telegrams	14.60	78.80

POSTPONING VISIT TWO OR THREE WEEKS OR WOULD YOU LIKE TO TAKE [NAME] IN MY PLACE REGRET UPSETTING PLANS TELEPHONE ME AT VERZ FOURTEEN MELBURY ROAD NINE OCLOCK THIS EVENING [NAME]

7.3.2 Salutation and Farewell

Whereas salutations are frequently omitted in text messages, they are even rarer in telegrams. Figure 7-2 shows that 96% of the telegrams do not contain any salutation formula. The reason for this phenomenon is cost: to save money, the sender does not begin the text with a greeting, but directly starts in medias res, i.e. with the actual message, as examples (232) to (234) illustrate.

(232) Arrive on Thursday morning Love [Name]
(233) Please wire me how mother is [Name]
(234) [NAME] DEAD OF HEART ATTACK TODAY. HAVE [NAME]. CAN'T KEEP HER PAST MONDAY. COME SUNDAY TRAIN. [NAME]

Figure 7-2: Salutations in Text Messages and Telegrams

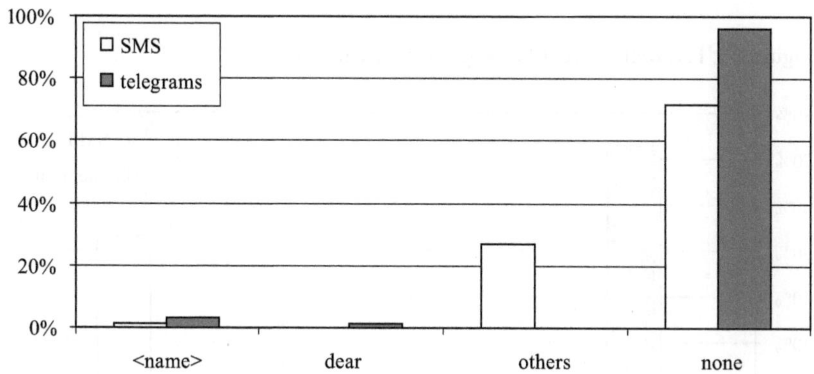

Salutation categories	% in SMS	% in telegrams
<name>	1%	3%
dear	0%	1%
others	27%	0%
none	72%	96%
total	100%	100%

This relation is different with farewells. Upon receiving and opening a text message, the recipient knows the sender of the message, for he can see his name or number on the display. For reasons of economy and efficiency, the sender's name is thus often omitted in text messages. In telegrams, however, which include no sender's address, it is important to add the name at the end of the message. Leave-taking formulas are thus not usually left out, unless the context makes it clear who the sender is.

Figure 7-3 indicates the percentage of farewells in telegrams compared to SMS and shows that telegrams are most often concluded by the sender's name (73%). Only 10% contain the formula *(lots of) love*, while 11% do not contain any farewell formula at all. The telegrams in quotations (235) and (236) present examples which include farewells. In quotation (237), in contrast, there is no leave-taking formula. Here, the recipient is able to guess the sender from the context so that no farewell is needed.

Figure 7-3: Farewells in Text Messages and Telegrams

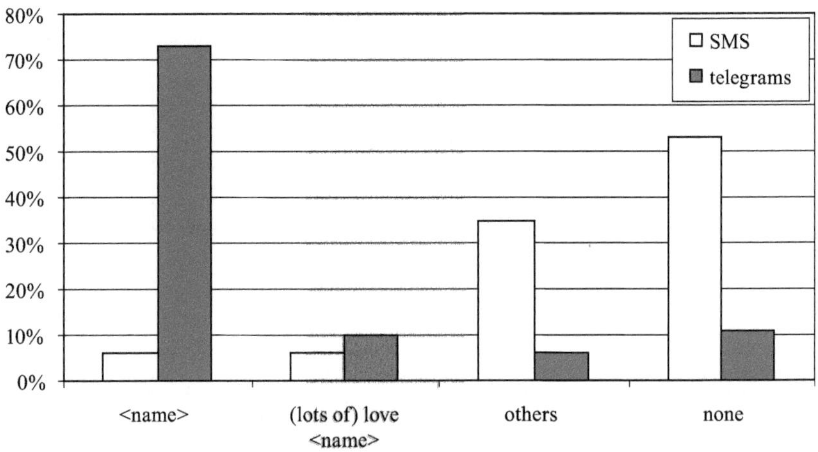

Farewell categories	% in SMS	% in telegrams
<name>	6%	73%
(lots of) love <name>	6%	10%
others	35%	6%
none	53%	11%
total	100%	100%

(235) BOTH COMING FRIDAY [NAME]. [NAME].

(236) HAPPY NEW YEAR DARLING MAY THIS ONE BRING ALL YOU DESIRE YOURS ALWAYS [NAME]

(237) Your letter received this morning crossing mine of last night. I am uncertain as to being in London on Tuesday. If Monday does not suit we might postpone to another early day to be fixed.

It can be concluded that there are significantly fewer salutation formulas in telegrams than in text messages, whereas there are notably more farewell formulas in telegrams than in text messages.

7.3.3 Grammatical Features

To keep the messages short and cheap, certain word-classes[6] have often been left out. Similar to text messages, it is mainly the subject pronoun *I* which is omitted: Figure 7-4 shows that subject deletion occurs 29.08 times per 1000 words in telegrams, which is 1.4 times more often than in text messages. Also more common is copula omission (13.13 times per 1000 words), followed by subject pronoun and auxiliary verb omission as well as article / possessive pronoun omission, both occurring 6.57 times per 1000 words. Quotings (238) to (241) provide an example each.

(238) Ø AM WELL AND FIT ALL MY LOVE DARLING KEEP SMILING [NAME]

(239) [Name] ø dead please come at once [Name]

(240) ØØ COMING HOME OK SHOULD ARRIVE SATURDAY SOMETIME [NAME]

6 Since telegrams frequently had an unusual typeface (cf. Section 7.1, Definition), they often did not contain any punctuation marks. This led to asyndetic sentences of the following kind: SORRY CAN'T COME HAVE LOST RATION BOOK [NAME]. It is thus impossible to analyse whether such a sentence merely lacks a full stop or whether it can be regarded as an asyndetic sentence which omits a causative subordination such as *because* or *as*. Therefore, asyndetic sentences are not examined in this corpus. Nevertheless, due to the lack of sentence-final punctuation marks, asyndetic sentences are rather frequent. Generally, the sentences are not complex and there are only a few conjunctions anyway: three coordinating conjunctions (*and* (2) and *or* (1)) as well as six subordinating conjunctions (*although* (1), *if* (4) and *as* (1)). In the whole corpus, only 18% of the sentences are marked by a full stop – 82%, by contrast, lack their sentence-final punctuation marks. What is more, there are no exclamation or questions marks in any of the analysed telegrams.

(241) BEWARE OF THE WIDOW HOPE YOU HAVE Ø GOOD TRIP LOVE [NAME]

Occasionally, but still more often than in text messages, prepositions as well as pronouns other than the subject pronoun are omitted (each of them 1.88 times per 1000 words). Generally, the tendency points towards even greater omission of words in telegrams than in text messages. This result might surprise. Yet, in contrast to SMS, there are only syntactic reductions and no lexical ones in telegrams. In telegrams, it is important to save words and not characters: it does not make a telegram cheaper if it contains abbreviated words, for the abbreviated word is as expensive as if unabbreviated. To keep a telegram cheap, it is therefore important to use as few words as possible. In text messages, however, one has to minimise the number of characters as well as the number of words. This is a crucial difference, which has led to different practices of economy. What is more, telegrams were not written (i.e. coded and decoded) by the senders themselves, but by official telegraphists who manually sent them via Morse signals. So, the composer of a telegram could not experiment with language, especially with spelling, which explains why telegrams neither make use of consonant spelling nor are there any apostrophe omissions or any other ways in which lexical items are abbreviated. The ludic moment inherent in text messages is thus largely absent in telegrams.

7.3.4 Lexical Features

Apart from a few spoken expressions such as the pet name *darling* (242) or the use of the dialectal degree modifier *pretty* (243), there are no other spoken features such as regionalisms, slang expression, discourse markers, or interjections in the telegram corpus. Spoken expressions occur 4.69 times per 1000 words.

(242) HAPPY XMAS *DARLING* ALL MY THOUGHTS ARE WITH YOU ALL MY LOVE IS YOURS [NAME] (My emphasis)

(243) CONGRATULATIONS AND GREETINGS FOR THE 21ST BIRTHDAY... FEELING *PRETTY* MISERABLE AT BEING IN BED... [NAME] (My emphasis)

Figure 7-4: Elliptic Phrases in Text Messages and Telegrams

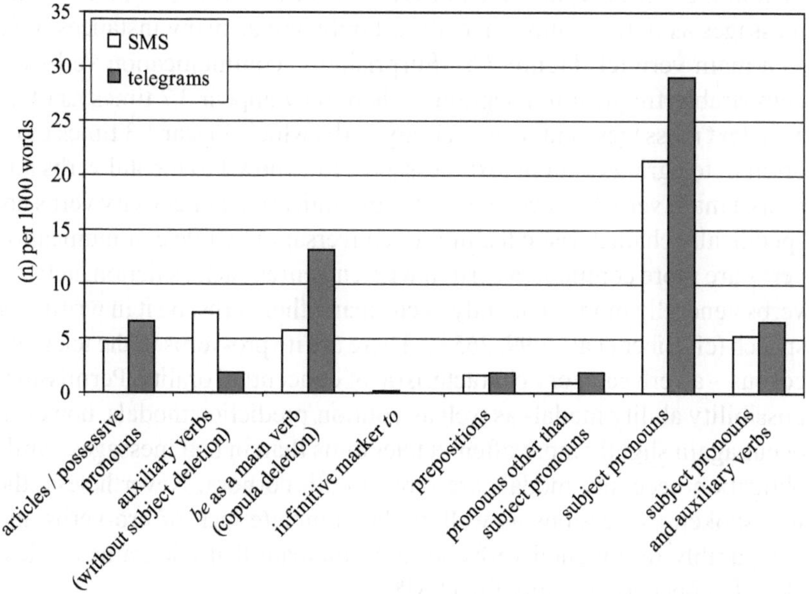

Ellipses	per 1000 words in SMS	per 1000 words in telegrams
articles / possessive pronouns	4.65	6.57
auxiliary verbs (without subject deletion)	7.36	1.88
be as a main verb (copula deletion)	5.76	13.13
infinitive marker *to*	0.17	0.00
prepositions	1.66	1.88
pronouns other than subject pronouns	0.94	1.88
subject pronouns	21.37	29.08
subject pronouns and auxiliary verbs	5.81	6.57
total	47.72	60.98

Similarly, there are significantly fewer mental verbs in the telegram corpus than in the SMS corpus – mental verbs occur 5.33 times as often in text messages as in telegrams – and almost only half as many instances of *be* as a main verb (cf. Figure 7-5). Surprisingly, communication verbs are comparably frequent in telegrams, where they appear 3.7 times as often as in text messages, and so are activity verbs which appear 1.3 times more often in telegrams than in text messages. In contrast to mental verbs and *be* as a main verb, however, neither communication nor activity verbs are specifically characteristic features of conversation: while communication verbs are more common in certain written genres such as fiction, activity verbs generally more frequently occur than other verbs, be it in writing or speech (cf. Biber et al. 1999: 265 f.). There are no pro-verbs in the telegram corpus – a verb category characteristic of conceptual orality. Permission/possibility/ability modals as well as volition/prediction modals, however, occur again slightly more often in telegrams than in text messages, while obligation/necessity modals are rare in both corpora. Nevertheless, the few spoken expressions as well as the complete lack of pro-verbs and comparably rare mental verbs strongly indicate that telegrams are less close to conceptual orality than SMS.

7.4 Telegrams in Koch and Oesterreicher's Extended Model

The linguistic analysis and comparison with text messages has shown that there are indeed certain parallels but also differences between telegrams and SMS. There are considerably more omissions of salutation formulas, while farewell formulas tend not to be left out in telegrams. Moreover, there are more syntactic reductions but no lexical ones, while spoken expressions, pro-verbs as well as mental verbs – all being typical of conceptual orality – are rare. The question now is where telegrams can be placed in Koch and Oesterreicher's extended model. From a medial point of view, they are clearly written, asynchronous and digital. They are medially digital, for communication via telegraphy physically

Figure 7-5: Verbs in Text Messages and Telegrams

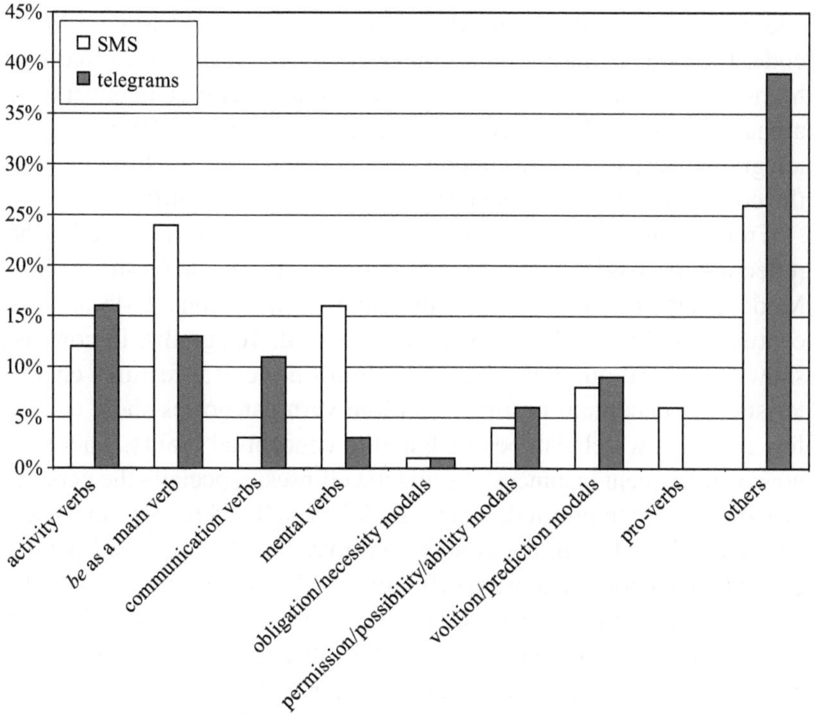

Verbs	% in SMS	% in telegrams
activity verbs	12%	16%
be as a main verb	24%	13%
communication verbs	3%	11%
mental verbs	16%	3%
obligation/necessity modals	1%	1%
permission/possibility/ability modals	4%	6%
volition/prediction modals	8%	9%
pro-verbs (e.g. *do, get, put*)	6%	0%
others	26%	39%
total	100%	100%

consisted of electrical impulses sent as equivalents of Morse code letters over firmly installed cables (cf. Schwitalla 2002: 34). In this respect, telegraphy is very similar to SMS: while the former is based on Morse code, the latter is based on a binary code[7] which is transmitted over high-frequency electro-magnetic waves. Each technique converts the alphabet into a code and decodes it again. The only difference is that telegrams were manually coded and decoded by telegraphists whereas the coding and decoding happens automatically with SMS.

From a conceptual aspect, the classification of telegrams needs more reflection. As has been shown in Section 6.3, Koch and Oesterreicher's Model Adapted to the New Communicative Conditions, SMS are conceptually oral as well as conceptually digital. Telegrams, in contrast, show fewer oral qualities: there are almost no lexical features characteristic of conceptual orality, while the verb categories show equally few elements which can be attributed to conceptual orality. There are, however, frequent instances of syntactic ellipses, especially the personal pronoun *I* is often omitted, which, as deictic self-reference, can usually be reconstructed from the context and may thus be neglected without causing major comprehension difficulties. All in all, it can be concluded that telegrams must be classified somewhere near conceptual literacy. The fact that punctuation marks are often absent, that telegrams may be composed in all upper case letters, the almost total absence of salutation formulas as well as the high amount of syntactic ellipses are further indications of conceptual digitality. In view of that, private telegrams (T_T) can be arranged in quadrant four in Koch and Oesterreicher's extended model. Figure 7-6 shows the model and the two text types, private text messages and telegrams, in relation to each other.

7 To get an idea of what a text looks like when converted into binary code, cf. any binary translator on the Internet. Binary translators provide a tool to convert regular text to binary and vice versa. The sentence 'I love you' looks as follows in binary code: 01001001 00100000 01101100 01101111 01110110 01100101 00100000 01111001 01101111 01110101. In Morse Code, the same sentence translates .. / .-.. --- ...- . / -.-- --- ..-

Figure 7-6: Telegrams in Koch and Oesterreicher's Extended Model

		conceptually oral ⟷ written	conceptually digital ⟷ analogue
medially oral analogue	synchronous		
medially oral digital	synchronous		
	asynchronous		
medially written analogue	asynchronous		
medially written digital	quasi-asynchronous		
	asynchronous	T_S T_T	T_S T_T

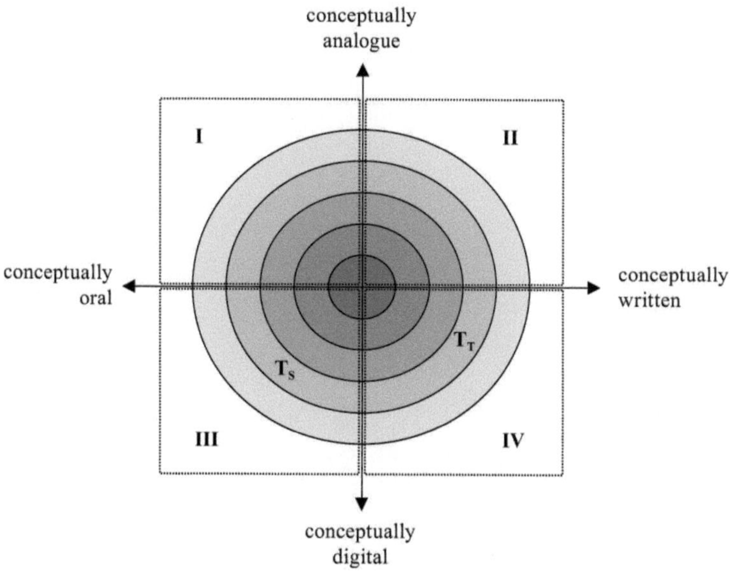

8 Multimedia Messages

8.1 Definition

MMS, short for Multimedia Message Service, is such a new service that the term is not even included in the *NetLingo Internet Dictionary*, published in 2002. As an enhanced transmission service and offspring[1] of the Short Message Service, it enables users to send messages with multimedia features such as still images (i.e. photos and pictures), sound (i.e. audio files) and, increasingly, also video clips via mobile phones. In contrast to SMS, MMS have medium media richness: "[t]hey have many of the same characteristics as SMS […] but the ability to send pictures makes them appropriate to use for communication where SMS does not offer sufficient richness" (Solberg 2004: 1).

Most camera phones support MMS, but there are also non-camera phones with which multimedia messages can be sent and received. In fact, "[m]ost of the first MMS-enabled phones that were introduced to the market just prior to and shortly after the introduction of MMS did not have integrated cameras" (Solberg 2004: 49). This, however, has changed so that nowadays "[a]lmost all new MMS-enabled mobile phones have integrated cameras" (Solberg 2004: 50). The term photo messaging is a closer description for taking pictures and transmitting them with mobile phones. If the recipient of a multimedia message does

1 "MMS, in part because of the similarity of the acronym that defines it, is naturally identified as the successor to the SMS (but forerunner of video communications). It is interpreted as a natural linguistic-expressive enhancement of SMS (we now have not only text, but also text and image). It is motivated by the same desire to conserve (and construct) one's own social memory. Moreover, the strong social entrenchment of the practice of sending SMS, the 'mythical' appeal based on the fact that the messages are experienced as absolute and unrepeatable, their emotional force and the specificity of their expressive form (i.e. the way they are composed) makes SMS a fundamental touchstone for MMS." (Colombo and Scifo 2005: 98)

not have an MMS-enabled mobile phone, he may be directed to a Web page where he can download the whole message, view the pictures and listen to the audio file.

An MMS can include more signs than a single SMS. Many mobile phones allow up to 1000 signs per message, as well as approximately 40 signs for a subject line. A feature which still does not work very well with the Multimedia Message Service is that of delivery reports. Even if mobile phones offer this function, it does not work properly with all providers. As with emails, which, provided they include the request for a reading confirmation, ask the addressee to click on the enter key to send a report, the recipient of a multimedia message may be informed that the sender has requested a delivery report and he can then decide himself whether he wants one to be sent or not. In other words, in contrast to short messages, it may be left up to the recipient to do so. Apart from these differences, MMS functions similar to SMS: like text messages, multimedia messages are of a dynamic nature and can be composed, put in drafts, edited, sent, archived, or deleted.

8.2 Historical Background

The first commercial camera phone was launched by J-Phone in Japan in October 2000 (Srivastava 2004: 17). Camera or picture phones, as they are also called, have been a big promise for mobile phone operators who expressed "high hopes that, following text messaging's success (especially in Europe and Asia), 'picture messaging' using camera phones will establish photos as a new genre in mobile communication" (Kindberg et al. 2005: 42). However, despite the optimistic forecasts, customers are reluctant when it comes to multimedia messaging. Results of surveys show that two thirds of the customers assert that they only want to telephone or send text messages and are not in want of any further or enhanced messaging services (Kuhn 2005: 3). Yet, this attitude should not be taken too seriously. In Section 3.2, Historical Background: SMS as Side-Product of Other Mobile Communication Services, it is pointed out that in its beginning in the 1990s, SMS was generally regarded as a gimmick by the majority of mobile phone users and almost a decade passed until the Short Message Service started to show an explosive increase in figures. A parallel process may be observed with photo messaging, which has only been introduced in Europe in 2002. Photo messaging is still in its infancy and it may similarly take its time to actually become popular among mobile phone users.

Whatever the future of multimedia messages looks like, the current situation is that "[w]ith the exception of Japan, mobile multimedia has not achieved the success of text messaging" (Koskinen 2005a: 3). Similarly, Orlowski (2004) finds that "[t]wo years after the introduction of picture messaging in Europe, MMS is still a flop". One of the reasons for this negative result certainly is the critical mass[2]: a study conducted by A.T. Kearney and the University of Cambridge shows that not even half of the mobile phone users possessed a camera phone in Western Europe in 2005 (Menon et al. 2005: 7). This means that even those who have a camera phone may only have little opportunity to use the Multimedia Message Service, depending on whether their addressees also own an MMS-enabled mobile phone or not. What is more, many people with camera phones either do not know about the function of picture messaging, or else have not activated it. Hence, the rise in picture messages has only been slow: the number of users that send or download multimedia messages in Western Europe increased only by 4% from 14% to 18% in the years 2004 and 2005, and among those who own a camera phone, 69% report that they never use the Multimedia Message Service, while 13% use it once per month and only 17% more regularly (Menon et al. 2005: 10).

A further reason for the slow growth of MMS certainly is cost. Whereas text messages have a fixed price, multimedia messages have a less clear price tag and are thus less appealing for users. Depending on the provider, multimedia messaging can be rather expensive, which "partly helps to explain why cheap but limited SMS is so popular while in many places MMS took a while to get going at all" (Battarbee and

2 The term "critical mass" is taken over from Markus (1987, 1990) who developed the concept of the critical mass in an email implementation pointing out "that it is not worth any individual switching from traditional media to email until a reasonable number of the people they want to communicate with are also using it" (Markus 1987, 1990 in Rudy 1996: 201).

Kurvinen 2005: 82). Usually, the cost of an MMS depends on its size. In other words, the sender of a multimedia message is charged for the amount of data that is sent, which leads to a further problem, namely that of picture quality. The better the picture quality is, the bigger its size is and the more expensive a message becomes. Also, pictures taken with a camera phone are still not always of a very good quality, which is another reason why people are reluctant to use MMS. Even worse, the photos networks deliver do not necessarily match the megapixel images that users think they are sending. Simon East, Cognima founder and former Psion software lynchpin and Symbian technology VP, explains that "[b]y default many phones send 30kb images, and you have to select a large image that's around 100kb manually [...] For example, the Nokia 7610 takes 1152x864 sized images, but even with the 'large' setting, a 576x432 sized image is delivered. You need at least 300kb to get something that's printable." (Orlowski 2004). Furthermore, "[w]ith the exception of Nokia 7650, the first MMS-enabled mobile phones all had relatively small displays" (Solberg 2004: 49), so that pictures could not be seen properly.

The critical mass, cost, picture quality and also the lack of reliability have retarded the successful take-off of multimedia messaging. Sending MMS nationwide is relatively reliable, but as soon as picture messages are dispatched from one country to another, delivery may fail, depending on the country and the roaming contracts between the mobile phone providers. It is thus still more reliable to send a postcard than an MMS from one's holidays abroad.

Besides technical and economic reasons, there are also social factors which help to elucidate why people do not use the Multimedia Message Service as often as SMS. Okabe (2005) finds that compared to text messages, multimedia messages may be perceived as intrusive and narcissistic if they contain a personal photo. For example, one of his informants remarked: "I can send a photo of me with a new hairstyle to my boyfriend, with the comment, 'How does this look?', but I can never send a picture like that to a friend. They'd think I was an idiot". So, photos are often rather a matter of personal use as figures of surveys show. Okabe (2004) states that 90% of the informants use their photos to "view them on their handset", while only 60% of the photos are actually shared via MMS. Kindberg et al. (2005: 48) similarly find that the "majority of sharing (one third of all images) took place face to face on the phone itself" while "[o]nly about one fifth of the images were shared by sending directly from phone to phone (largely via MMS)". In the light of these results, camera phones have even been characterised as a "capture and show" rather than a "capture and send" device (cf. Kindberg et al. 2004: 12; cf. also Vincent 2004: 8). This preliminary conclusion, however, may be too vague, particularly with regard to the fact that picture phones are still in their infancy and their full potential has by far not been tapped yet.

8.3 Structure

Basically, the structure of a multimedia message is similar to that of a text message. Its pattern is fixed; the only difference is that it enables its users to add images or audio files to their messages. Some mobile phones further have the option to insert a subject. Depending on the make and model of the mobile phone, the subject line is limited to about 40 signs. If the text is short, it may happen that the subject line is used for the whole message, there being no text in the body at all. The multimedia message in (244) provides an example: here, the whole message is written into the subject line.

(244)	Subject Line	Image	Body	Audio File
	Guetnacht, min ängel!!![3]		[none]	[none]

3 Goodnight, my angel!!!

Similarly, the multimedia messages in example (245) contain written messages in the subject line only. The sender of the MMS was spending some time in Stein am Rhein, a small town in the North East of Switzerland, and upon passing its picturesque Old Town, he took some photos and sent them to his addressee who could then get a visual impression of the town. The subject line, or at least part of it, can be read even before opening the message. It usually appears below the signature, i.e. below the name or number of the sender, and informs about the content before the message is opened.

(245)	Subject Line	Image	Body	Audio File
	schös altstädtli [4]		[none]	[none]
	Mit erkerli, hehe[5]		[none]	[none]

Like SMS, MMS can also be linked, which means that it can contain several pages. If the composer of a multimedia message wants to add a second picture to his message, for instance, he can simply add a further page, select the picture and attach it with or without text. Newer mobile phones even support SMIL. SMIL, short for Synchronised Multimedia Integration Language, is a markup language which combines text with sounds, images, videos, etc. and sets them up as multimedia presentations. With this feature, various parts of a multimedia message can be

4 beautiful little old town

5 With little bay windows, hehe

arranged into small multimedia slideshows. Provided that the recipient has a SMIL-enabled mobile phone, he can then view the MMS as a slideshow on his mobile phone. Yet, there is no such example in the analysed corpora, as MMS-slideshows are even less widespread than single multimedia messages with one picture/photo only and often, users do not know about this function at all. Accordingly, the multimedia messages in example (245) could have been sent as one message consisting of two pages if the sender had been aware of the SMIL function in his mobile phone. As he did not know about this feature, however, he sent two separate messages.

Even more than SMS, MMS can be compared to postcards: the parallel with MMS is that we have the possibility to send a picture from one's holiday resort, for instance, accompanied by a text to a recipient specified by the sender. Example (246) presents such a postcard-like multimedia message from some holidays in Spain.

(246)	Subject Line	Image	Body	Audio File
	[none]		He [Name], Grüße aus Spanien, [Name] und ich vermissen dich, du fehlst uns sooooo sehr (besonders für dieses bescheuerte Photo ;o)[6]	[none]

If the Multimedia Message Service was more reliable, there would certainly be more postcard-like multimedia messages sent from holidays abroad than statistics show. Yet, MMS from abroad often never reach

6 Hey [Name], greetings from Spain, [Name] and I miss you, we are missing you sooooo much (especially for this daft photo ;o)

their recipients and get lost forever somewhere between the provider of the sender and that of the recipient. What is more, multimedia messages are relatively costly from abroad, so that people rather send a text message greeting or else a traditional paper-based postcard. Once MMS becomes more reliable and cheaper, however, the speculation is that it will replace postcards, for the main difference and its advantage is transmission speed (cf. Koskinen et al. 2005: 60). The multimedia message in example (246), for instance, reached its addressee within a few minutes, whereas it would have taken several days for a postcard to arrive.

8.4 Rules and Conventions

Principally, the general rules of netiquette and textiquette apply to photo messaging, too, whereas a special MMSiquette does not yet exist. Vincent (2004: 7) pinpoints it well when she broaches the issue of the etiquette for camera phones:

> The dilemma for the user of the camera phone is that, although there is an etiquette for using cameras in public places, there is not one for sending the image to a third party without the consent of the person you have captured in the photo. Etiquette exists for the use of talk and text but the etiquette for camera and for the transmission of data […] has yet to be confirmed.

Some versions of camera phone etiquette can certainly be found on the Internet, but there are no generally binding rules with which the majority of users are familiar. One of the rules when it comes to photo messaging is that one should "[r]espect other people's privacy" and "[s]ave faces" (Avec Mobile 2004). However, example (247) proves that people do not always follow this rule. Here, the sender took a photo of someone with whom he was sharing the compartment in a train. The person made a strange impression on the sender simply for the fact that he was eating a tomato and reading Kafka. Nevertheless, he thought the situation special enough so as to share it with a friend of his.

(247)	Subject Line	Image	Body	Audio File
	[none]		Hehe, e kurrligi gstalt sitzt do vis à vis vo mir: isst e tomate und liest kafka…!![7]	[none]

There seems to be some uncertainty about what the appropriate photo messaging behaviour might be. Questions of ethics arise and moral issues have to be considered. Moreover, one should be aware of the fact that as soon as a picture is sent off, it has the potential to become public. One never knows what a third party will do with the message and should thus be sensitive and considerate when sending pictures of people other than oneself.

8.5 The Proportion of Image to Text

With MMS-enabled mobile phones and especially with camera phones, which make it possible to catch a particular moment visually and directly send it to other people, image communication receives a new quality. A medium which not only allows the immediate production of pictures, but also offers the option to send them right away changes the way in which people communicate. It provides "a tool for sharing that very same space and objects that may be there" (Scifo 2005: 369), thereby including absent people and making them part of the situation. In multimedia messages, pictures and text are inseparably intertwined. This means that pictures are an essential part of multimedia communication so that the combination of image and writing reaches a new dimension.

7 Hehe, a quaint figure is sitting opposite of me: is eating a tomato and reading kafka…!!

Dialogic distance-communication has so far been detached from the visual channel. Images, of course, can also be enclosed to letters, be faxed or attached in emails, i.e. it is certainly not a new achievement to send images. What is new, however, is the immediacy of taking pictures and sending them. A medium which makes it possible for everyone to catch pictures and dispatch them via the same medium to other people right afterwards is a new technical step in communication which brings about a new understanding of image communication. A good example is the following sequence of multimedia messages which is presented in (248).

(248)	Subject Line	Image	Body	Audio File
	[none]		Tramhaltestell platte, usstiege, am careum verbi, trottoir entlang de tramschiene[8]	[none]
	[none]		Do isch au scho dr häldeliweg, gad neb dr bustell[9]	[none]
	[none]		Stegeli druf...[10]	[none]

8 Tram stop platte, get off, pass the careum, pavement along tram tracks

9 Häldeliweg is already here, right next to the roadworks

10 Go up the steps...

[none]		Innegoh und die erschti tür links... houououou... beschwörigstanz für [name][11]	[none]

Here, the sender of the messages documents his instructions how to get to a lecture room. The composer of these messages, a student from Zurich University, was asked where lecture room E3 at Häldeliweg 2 is, in which some exams were taking place. The answer not only consisted of a written description, but also of a visual illustration, explaining the way from the tram stop to the lecture room. This example proves well how the use of images changes text. The deictic spatial expression *Do* (meaning 'here') in the second message would have no meaning without the corresponding picture nor would the instruction *Stegeli druf...* in the third message have any value of truth if there was no image of the steps attached. In other words, without the photos, the description of how to get to the lecture room would have had to be more precise and less deictic. By attaching the photos, however, much of the verbal description can be shortened in favour of the images, for the reference object is thus made present.

The multimedia message in (249) provides another example in which the proportion of image and text is such that the photo replaces a more exact description of the item. In fact, any verbal portrayal of the ski pants would never have been as precise as a photo of them. The person who tries them takes a photo of herself while holding the camera towards the mirror of the changing room and then sends it to her partner, wondering if the ski pants will also be liked by him. At the same time, she cannot make up her mind regarding the colour of the pants and asks whether she should buy them in white or in black.

11 Enter and the first door at the left-hand side... houououou... magic dance for [name]

(249)	Subject Line	Image	Body	Audio File
	[none]		Gad am skihösi probiere, grins... gfalled sie dir? sölli sie in wiis oder in schwarz neh?[12]	[none]

Kindberg et al. (2005: 48) find that "the spontaneous capture of visual information to help achieve task", such as showing some clothes which are tried on, constitutes a new form of communication. And they conclude that "[w]hen information is fundamentally visual in nature, neither text nor voice will suffice". Similarly, in the multimedia message in (250), the sender asks whether the chosen clothes would be appropriate or too smart for the upcoming gathering. Here, too, a more precise description is not necessary, as the photo shows all the relevant details.

Examples (249) and (250) illustrate well that by combining images with text, things can be communicated which are beyond description. The senders could never have described their clothes with such accuracy as to give their recipients a clear idea of them. By including a photo, however, the addressee gets an exact picture of the situation and can thus be of help.

(250)	Subject Line	Image	Body	Audio File
	[none]	*a photo of the sender dressed in a top, skirt and boots*[13]	Würde das mit den stiefeln gehen? oder ist das zu schick? turnschuhe passen nur zur hose, hab keine saubere mehr. kann evtl noch ein anderes oberteil mitnehmen.[14]	[none]

12 Just trying a pair of ski pants, grin... do you like them? shall i take them in white or in black?

13 For reasons of privacy, the picture cannot be printed here.

14 Would that go with boots? or is it too chic? sneakers only go with trousers, i haven't got any clean ones. may also take another top with me.

> The combination of camera and direct sending capabilities lets us use images to bring remote people into an experience or to accomplish tasks with them. This is distinct from related messaging activities in several ways. (Kindberg et al. 2005: 48)

Unlike images emailed later, the photos in a multimedia message can be captured and shared at the same moment. Unlike text messaging, many images sent to an absent person are visual evidence of an event. Often, such messages are sent with little or no need for further explanation, as they make sense because of the shared context and understanding, which, of course, depends on the closeness of a relationship (cf. Kindberg et al. 2005: 48). Accordingly, the multimedia message in example (251) does not make much sense to someone who is not familiar with the context. For the recipient, however, the message was highly meaningful: the person who took the picture was attending a wedding with his music group and while they were holding their heart-shaped balloons to be released into the air, he took a picture of them and dispatched it right away to share the moment with his addressee. Similarly, Mäkelä et al. (2000: 553) point out that

> images are only useful if the receiver has a clear understanding of the sender's daily life and the context in which the image was sent. To be sure: An image can contain much more information than a voice or text message. Nevertheless, each message must be translated by the receiver according to certain rules. In the case of text or voice messages the receiver's "linguistic competence"[15] allows them to give meaning to other's "linguistic performance". In the case of image messages senders and receivers have to produce rules of their own. However, that also means that people not belonging to the same peer group have no possibility to understand image messages or to send ones.

(251)	Subject Line	Image	Body	Audio File
	Ballön[16]		[none]	[none]

15 Chomsky (1957) in Mäkelä (2000: 553).
16 Balloons

The fact that much is said through the attached photos or images in multimedia messages leads one to conclude that the average text-length is not much longer than in text messages. Theoretically, an MMS, which provides up to 1000 signs, could be more than six times as long as an SMS, which consists of maximally 160 signs. Figure 8-1, however, indicates that multimedia messages tend to be even shorter than text messages[17]. Only the German MMS are slightly longer both in terms of words and characters than the German SMS. Swiss-German MMS are generally shorter and so are the English ones, which show a considerably smaller number of words and characters than text messages.[18]

Figure 8-1: The Length of Multimedia Messages and Text Messages

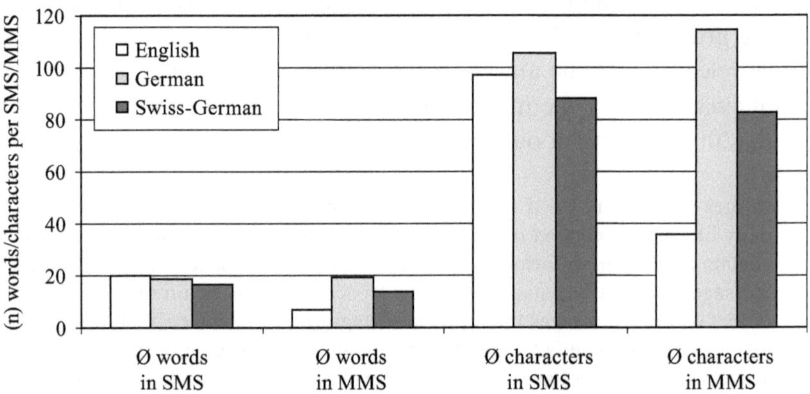

Length	Ø words in SMS	Ø words in MMS	Ø characters in SMS	Ø characters in MMS
English	19.90	6.74	96.97	36.15
German	18.35	19.05	105.37	114.16
Swiss-German	16.77	13.80	88.11	82.95

17 The statistics also include the words in the subject line, for 14% of all the MMS in the corpora have some text in the subject line only.

18 Since the German and English MMS corpora are both relatively small, their results may not be too indicative. Figure 1-2 and 1-3 in Section 1.3, Methodology: Data Collection and Corpora, show that the mean length of SMS is equal throughout all

7.5% of all the multimedia messages in the corpora even contain an image only without any text at all. An example of such a commentless MMS is provided in (252). Here, the sender has just received his new wooden cupboard and sends some impressions of it to his friend to share his delight. Whereas the first two MMS contain a small message in the subject line, the third MMS is blank and only has a photo attached.

(252)	Subject Line	Image	Body	Audio File
	Schös Schränkli, gell?[19]		[none]	[none]
	Herzigi schlößli![20]		[none]	[none]
	[none]		[none]	[none]

corpora. This finding also applies to emails (cf. Figure 1-1). It can thus be hypothesised that the length of MMS would be more balanced if the German and English MMS corpora had been more substantial. Taking the more comprehensive Swiss-German corpus for reference, it can be concluded that the texts in MMS tend to be slightly shorter than in SMS.

19 Beautiful cupboard, isn't it?

20 Sweet locks!

8.6 Image Categories

Scifo (2005: 367 f.) remarks that "communicating through images is closely and almost exclusively linked to one's network of strong relationships (friends from the same gang, intimate and/or long-standing friendships, girlfriends/boyfriends or wives/husbands, brothers, and to a slightly lesser extent, colleagues and schoolmates)". Similarly, Okabe (2004: 16) finds that "[t]hrough the capture and sharing of small but significant visual events in people's lives, camera phones are contributing to a kind of everyday photojournalism, where people are attentive to images and events that might be interesting or newsworthy events." Most of these photos, however, "are trafficked among peers, and are newsworthy only among friends and families". Döring et al. (2005) likewise assert that MMS are largely sent between people among whom a high level of trust has already been established (Döring et al. 2005 in Light 2005). Generally, as multimedia messages can be intrusive, "[d]ecisions of whether to send a photo or what kind of photo to send are made based on social relationships" (Okabe 2005). Pictures of a more intimate character such as the one in (253) are thus usually shared with close friends only.

(253)	Subject Line	Image	Body	Audio File
	[none]		my first ever bachelors meal lol :-)	[none]

In fact, the dinner, which is depicted in (253), is unspectacular: it consists of a soup, toast, chocolate, and tea. Nevertheless, the sender regards the moment as special: it is his first evening in his new flat and he wishes to share it with someone who is close to him. He takes a picture of his meal and sends it via his mobile phone, thus giving a relatively banal occurrence extra value. "[T]he mundane", as Ito and Okabe (2003) call it, is thereby "elevated to a photographic object", while everyday reality becomes "the site of potential news and visual archiving". Such a meal would never have been simultaneously shared with others before the advent of camera phones. They "provide a new tool for making these everyday *neta*[21] not just verbally but also visually shareable" (Okabe 2004: 13). "What counts as newsworthy, noteworthy and photo-worthy spans a broad spectrum from personally noteworthy moments that are never shared [...] to intimately newsworthy moments to be shared with a spouse or lover [...]" (Ito and Okabe 2003). Koskinen (2004: 11) points out that "senders have to *make* drama out of the banalities of ordinary life". Such a banal moment is also captured in example (254).

(254)	Subject Line	Image	Body	Audio File
	[none]		...Of em wäg is hotel krone, lenzburg ;-) hihihi[22]	[none]

Here, the sender is on the way to the Hotel Krone in Lenzburg, where he is going to celebrate his sister's birthday. It is a normal September evening and there is nothing off the beaten track. However, the sender regards the situation as important enough as to share it with his addressee. Similarly, the multimedia messages in example (255) show how a simple moment can almost be stylised to a unique occasion. The sender takes a photo of his mini-pizzas that are cooking in the oven and later takes a picture of his full stomach while explaining that he has

21 The first camera phone was launched in Japan (cf. Section 8.2, Historical Background). "In Japanese, 'material' for news and stories is called 'neta'. The term has strong journalistic associations, but also gets used to describe material that can become the topic of conversation among friends and family: a new store seen on the way to work; a cousin who just dropped out of high school; a funny story heard on the radio." (Okabe 2004: 13)

22 ...On the way to hotel krone, lenzburg ;-) hihihi

eaten all of them. Scifo (2005: 368) concludes that "[u]sually, sending an MMS message is meant to give access to and to share the place in which the subject is, or his private sphere". The purpose of such photo messages is thus simply to share one's privacy and let someone be part of one's personal life. The distance between the sender and receiver is diminished and for one moment, the sender becomes present for the addressee.

(255)	Subject Line	Image	Body	Audio File
	Brutzel, brutzel...[23]		das ischt eine feine französische pizza provencale... mjam![24]	[none]
	Sönd öllö dö drön...[25]	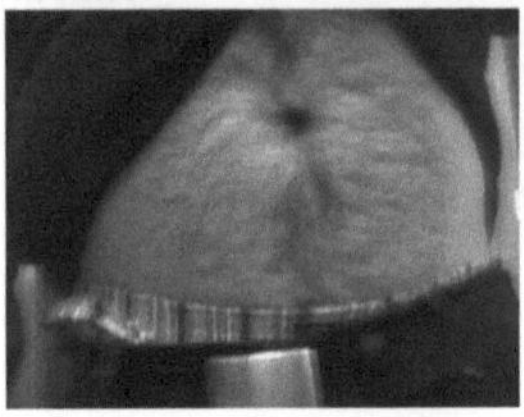	[none]	[none]

Okabe (2004: 7) explains that the photos "are not posed or staged, or particularly well-framed or thought out. Rather, they are snapped casually, with the intention of possibly looking at them a little later, recording a momentary slice of a viewpoint of everyday life". This is also well illustrated with the messages in example (255): the pictures are taken on the spur of the moment without being carefully prepared.

Besides visual contents, multimedia messages can also have audio contents. The practice of capturing and combining sound clips with photographs has been called "audiophotography" by Frohlich and Tallyn

23 Sizzle, sizzle...

24 this is a good french pizza provencale... yummie!

25 They are all in here...

(1999: 296). Frohlich and Tallyn (1999: 296 f.) find that audio augments photos in many ways, explaining that "audio might be of particular benefit, not just for adding voice annotations to photos [...] but also for capturing ambient sounds occurring around the time photos are taken". According to their study, sounds may enliven photographs and "bring static photos to life". An example is provided in (256). Here, the sender has captured a photo of some chickens on the road and attaches a voice record, imitating their cackle. Photographs, however, may equally enliven sound if an audioworthy sound is recorded and then illustrated by an additional photo. To that effect, the multimedia message in example (257) offers the addressee an audio test of the ongoing music rehearsal recorded while the people on the photo were playing their instruments. The audio test is then illustrated by a photograph so as to give the recipient not only an audial but also a visual impression of the neta.

(256)	Subject Line	Image	Body	Audio File
	put, put, put[26]		put, put, put... hüehndli unterwegs![27]	voice record: imitated cackle

(257)	Subject Line	Image	Body	Audio File
	Posthorn in a, menuett, w.a.mozart[28]		musikalische grueß usm workshop[29]	audio test of the minuet

26 put, put, put

27 put, put, put... chickens on the way!

28 Posthorn in A, minuet, w.a.mozart

29 musical greetings from the workshop

Figure 8-2: Image Categories in Multimedia Messages

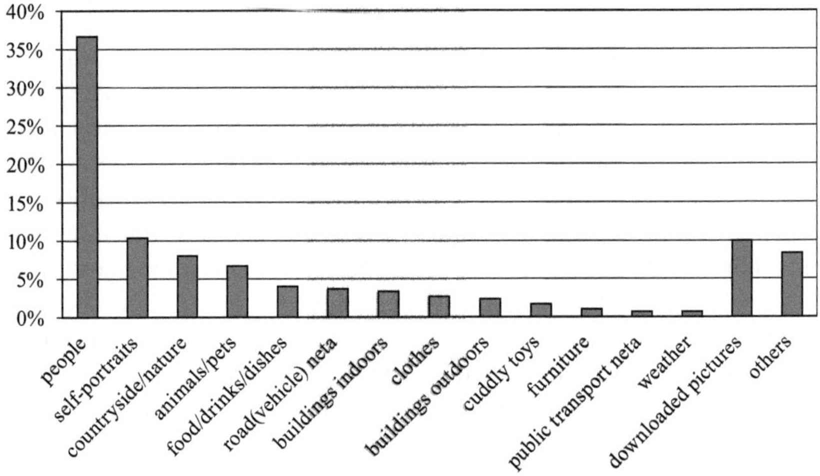

Examples (253) to (257) have in common that they all contain situations which could not have been shared before. They seem to enable "the *pluralization* of places (and experiences) and their *trans-location*" (Scifo 2005: 365). Pictures, of course, can also be taken with traditional cameras, but camera phones occupy a unique niche:

> In comparison to the traditional camera, which gets trotted out for special excursions and events – noteworthy moments bracketed off from the mundane – camera phones capture the more fleeting and unexpected moments of surprise, beauty and adoration in the everyday. They also invite sharing that is more immediate, ad hoc and ongoing. (Ito and Okabe 2003)

Often, the pictures are very spontaneously taken and so, images such as a trivial black board or a copy machine, a letter box or a snapshot of some animal is captured and shared with others. Figure 8-2 provides a synopsis of the most common objects and shows that most photos are taken of other people (37%) or of the sender himself (10%), a result similar to that of Kindbert et al. (2004: 5) who find that roughly half of the pictures are images of people. They are followed by photos depicting some form of countryside/nature (8%) and animals/pets (7%). Less frequent are pictures of food/drinks/dishes (4%), road(vehicle) neta (4%), buildings indoors (3%), clothes (3%), buildings outdoors (2%), cuddly toys (2%), furniture (1%), public transport neta (1%), and the weather (1%). Generally, it can be concluded that "[p]eople share images of familiar objects and people, private life (like objects, relatives), and social networks" (Koskinen and Kurvinen 2005: 61).

Messages may also contain downloaded pictures (10%), especially if they are used for special occasions such as wishing someone good luck for an exam or sending someone birthday greetings or New Year's wishes. Example (258) shows a message dispatched during the FIFA World Cup 2006. After the qualification of the Swiss for the last sixteen the following multimedia message containing the Swiss flag with an audio file of the national anthem was circulated; the subject line reads "Hopp Schwiiz".

(258)	Subject Line	Image	Body	Audio File
	Hopp Schwiiz!![30]		[none]	Swiss National Anthem

Yet, multimedia messages are not always welcome by those who receive them, as Koskinen (2004: 11) points out. To catch the recipient's interest, senders may use *"interest arousers"* (Sacks 1994, II quoted from Koskinen 2004: 11) and send something "which may be characterised as funny, uncommon or shocking enough to make them worth sending" (Koskinen 2004: 11). Example (259), for instance, is of such a nature.

30 Go Switzerland go!!

It shows a black photo taken in the dark on which nothing can be seen, while example (260) depicts a photo of a pigeon[31] in a tube carriage.

(259)	Subject Line	Image	Body	Audio File
	Häsi by night ;-)[32]		[none]	[none]

(260)	Subject Line	Image	Body	Audio File
	[none]	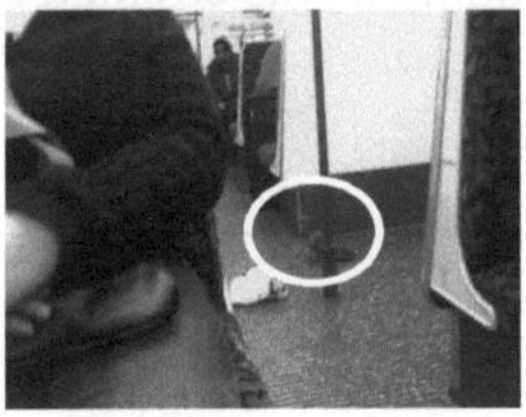	This little pigeon rode 2 stops on the tube haha!	[none]

However, "to guarantee recipient's attention, senders may resort to not just interest arousers, but to *'response-prompting-actions'*" (Koskinen 2004: 11). As best examples for such response-prompting-actions, Koskinen names questions and riddles. They force the recipient to respond in some way and react to the message. The multimedia message in (261) provides an example of such a riddle: the sender has covered a small present which he is going to give the recipient later that day; he takes a picture of it and sends it in the hope to get some answer back.

31 For reasons of clearness, the pigeon is circled. There is no circle in the original photo.

32 Bunny by night ;-)

(261)	Subject Line	Image	Body	Audio File
	Dis gschänkli...[33]		[none]	[none]

A similar riddle is provided by example (262). Here, the photo depicts the small cover of a Samsung mobile phone. This little piece of 1.5 x 0.5 centimeters has to be taken away whenever the mobile phone needs charging. One day, it was lost and could not be found anymore. It was only weeks later, when it unexpectedly appeared amidst some pens in a pencase. Upon its detection, the sender immediately took a picture of it and sent it as a multimedia message to the person who lost it, asking what this little piece might be.

(262)	Subject Line	Image	Body	Audio File
	[none]	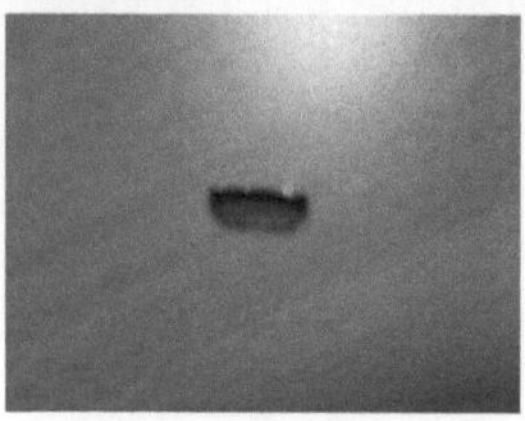	Was esch das?[34]	[none]

Koskinen (2004: 12) explains that riddles "typically consist of three phases, the riddle-initiating question, guesses, and the right answer that closes the riddle". Koskinen, however, gathers his data from design studies such as the *Mobile Image* and *Radiolinja MMS Study* projects[35].

33 Your present...

34 What is this?

35 The two projects, *Mobile Image* and *Radiolinja MMS Study*, were conducted in Helsinki, Finland, from 1999 to 2003. They were carried out by the industrial designers

In the *Mobile Image* project, informants were given a Nokia 9110 mobile phone and a Casio digital camera connected with an infrared link, while informants were equipped with an MMS phone in the *Radiolinja* study. In both projects, participants knew that they were being studied, which, of course, had a certain influence on the nature of the collected multimedia messages (observer's paradox). What is more, the phone service was free of charge, i.e. informants did not have to pay for their messages so that cost was no limiting factor. People could freely experiment with MMS (Koskinen 2004: 7 f.). They had to make use of the Multimedia Message Service as they took part in a comprehensive research project and had to contribute to the MMS data. It hardly needs mentioning that such a setting does not represent reality. As pointed out above, people are generally reluctant when it comes to multimedia messaging, and so it is not surprising that there is no such riddle-pattern found in the corpora that have been analysed for this study. The answers on the above riddles came via a telephone call (261) or SMS (262), with no further multimedia messages sent back and fro.

Esko Kurvinen and Katja Battarbee as well as by several assistants. Their focus lies on how mobile phones are used in sending and receiving digital messages and on MMS (cf. Koskinen 2004: 7 f.).

8.7 Multimedia Message Dialogues

There are not many MMS dialogues in the corpus. If multimedia messages are sent to and fro, they usually consist of two turns. An example of such a two-turn dialogue is provided in (263).

(263)	Subject Line	Image	Body	Audio File
	[none]		Hallo ihr zwei!hoffä ihr gnüsseds chli in davos und händ au so schöns wetter wiä mir!nebet bi häts no es bild vom aktuelä stand vo dä bauwiesä,ihr gsehnd es gahd vorwärts,also händ kei sorg und gnüssed i vollä züg euri feriä! Liäbs gruässli [name][36]	[none]
	[none]		Hallo [Name] Da lauft ja alles wie am schnürli. Liebi grüessli us davos.[37]	[none]

Here, the sender initiates the dialogue by sending a picture of the status quo of some building sites to some friends who spend their holidays in Davos, a holiday resort in the Grisons/Switzerland. In the message, the

36 Hello you two!hope you enjoy it a little bit in davos and also have such beautiful weather as we have!attached is a picture of the status quo of the building site,as you see,things are progressing,so don't worry and enjoy your holidays to the full! love [name]

37 Hello [name] everything seems to be going like clockwork. Love from davos.

sender further indirectly inquires about the weather and as a proof of their equally sunny weather, the recipient answers with a photo which provides a visual impression of the cloudless sky. In fact, Kindberg et al. (2004: 12) find that in contrast to text messages, many picture messages can be regarded as visual evidence or proof of something which has occurred. Similarly, in example (264), the sender has taken a photo of an ongoing thunder storm and shares it with a friend who likewise answers with a photo, which, however, depicts cloudy but otherwise nice spring weather. And in example (265), the sender shares her delight of a passing zeppelin and finally receives a picture of a balloon in the sky.

(264)	Subject Line	Image	Body	Audio File
	gwitter[38]		wow, dr himmel isch rabeschwarz!![39]	roll of thunder recorded by the sender
	Bi eus eschs ganz gmüetlech...[40]		[none]	[none]

38 thunder storm
39 wow, the sky is pitchblack!!
40 It is very cosy here...

(265)	Subject Line	Image	Body	Audio File
	Lueg emal, en zeppelin flügt gad verbii[41]		[none]	[none]
	Do flüget d ballön!![42]		[none]	[none]

Yet, the answer to a multimedia message does not necessarily have to be another multimedia message. It can as well be a text message or a telephone call to thank for the picture, which may be regarded as a small gift from the sender: it shows the sender's willingness to share his neta with the recipient or simply constitutes a sign of attention. In other words, "[p]eople share things to delight recipients, who are obliged to return the gift and show gratitude, or face sanctions if they do not reciprocate" (Koskinen and Kurvinen 2005: 61). Multimedia messages, however, may be too slow to fit the immediacy of a response. Thus people rather answer via another channel for "[d]epending on the elements that are inserted into the message it can be either faster (i.e. more synchronous) or slower (i.e. less synchronous) to respond than it is with an SMS. If just a picture is inserted and sent as a reply it is faster / as fast as responding with a short text message (SMS), however, if the message is created with several elements it can be slower than responding with an SMS (depending on the length of a text message)" (Solberg 2004: 103).

The only multimedia message dialogue in the corpus consisting of more than two turns is the one in example (266). In fact, it can in the

41 Look, a zeppelin is just passing

42 Here, ballons are flying!!

broadest sense be regarded as a riddle-like dialogue which the sender initiates by sending an outlook from the bus stop. The second message conveys a photo of feet standing somewhere. The next message is an answer from the recipient who is on the motorway and has taken a picture of what he sees, while the reply is a view of some underground shops. The enclosing text promises that the sender will get in touch soon (the implication is that the sender will do so via another medium, presumably calling the recipient on his mobile phone). Before, however, a last message is sent, showing that the sender, who still finds himself on the motorway, has now driven into a tunnel.

(266)	Subject Line	Image	Body	Audio File
	Standort[43]		Hihi, do standi jetzt gad, smile ui, bus chunt![44]	[none]
	Und jetzt da...[45]		hehe... und jetzt bini gad da gstande, grins...[46]	[none]
	ich bin da[47]		und ich bin da... hehe, nüme normal *g*[48]	[none]

43 Position
44 Hehe, i'm standing right here now, smile oh, bus is coming!
45 And now here...
46 hehe... and i've just been standing here, grin...
47 i'm here
48 and i'm here... hehe, not normal *g*

hehe[49]		und ich jetzt da... meld mi gad!![50]	[none]
ha voll de Tunnelblick... hehe[51]		[none]	[none]

Typically, however, an MMS dialogue consists of two turns of the kinds seen in examples (263) to (265). If some information has to be exchanged quickly, the composition of a multimedia message may simply be too laborious and transmission not fast enough. It then is quicker and cheaper to compose a text message, so that the dialogue is continued on the level of text messaging rather than multimedia messaging. Also, there is not always a neta at hand to be photographed, so that the recipient may choose to respond via another channel than that of MMS. What is more, the content of multimedia messages is often of such a nature that it does not trigger a longer dialogue.

Scifo (2005: 371) distinguishes three visual messages: performative, informative and problem-solving multimedia messages. A performative MMS is "employed as a resource to generate an act". Such a message would be one as shown in example (267). Here, the photo depicts the widely open mouth of the sender and the text clarifies its implication: the sender is hungry and wants the recipient to come home.

49 hehe

50 and i'm now here... i'll get in touch in a minute!!

51 i entirely have the view of a tunnel... hehe

(267)	Subject Line	Image	Body	Audio File
	[none]	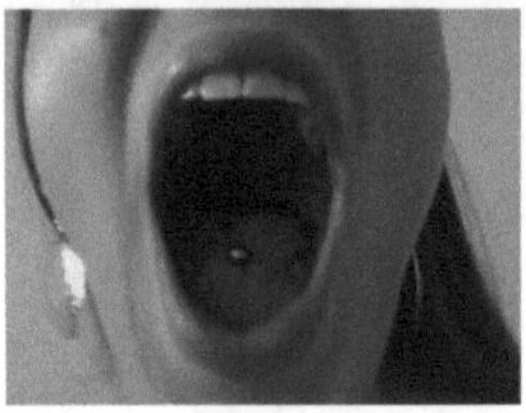	Wänn chunsch hei? Ich ha Huuuunger!![52]	[none]

In informative MMS, "visual communication [is] dominated by an informative function" (Scifo 2005: 371). Such an example is provided in (268). The sender of the message has been asked to post some letters on the way home and to make sure not to forget it. To prove that the letters have indeed been posted, a photo is caught of the very moment of throwing them into the letter box. The picture is then sent to the addressee to attest that the task has been fulfilled.

(268)	Subject Line	Image	Body	Audio File
	;-)		ha d briefe gad iigworfe…[53]	[none]

Problem-solving MMS contain "instrumental, pragmatic photos, taken in order to reduce time and cost, or to solve emergencies" (Scifo 2005: 371). A typical example is the one in example (249), where a pair of ski pants are tried on. Before they are actually bought, they are visually recorded and submitted to the sender's partner who is asked for his

52 When are you coming home? I'm huuuungry!!

53 i've just posted the letters…

opinion. Another example of a problem-solving multimedia message is the one in example (250) above.

None of these examples of performative, informative or problem-solving multimedia messages, however, have led to a dialogue. Whereas a text message is more likely to set off a dialogue, this seems to be different with MMS. In photo messages, the visual component plays a superior role. The aim is to visually communicate something which may not be described by words alone. Once the picture has been sent, the actual goal is achieved: what needs to be said (or rather: what needs to be shown) has been communicated. It is a singular act. Consequently, people are not likely to get involved in phatic communion with MMS. What is more, pictures are not needed for phatic communion or dialogues which, for instance, discuss an appointment or any other informative, performative or problem-solving topic. Accordingly, Solberg (2004: 103 f.) explains that "while an SMS usually is answered by an SMS, an MMS message is often answered by an SMS message [...] for MMS it seems that the answer usually does not need to contain more than text. In some cases, however, an MMS is answered with a richer medium, e.g. a phone conversation, if this is necessary in order to reduce the equivocality".

9 Linguistic Change and Possible Implications of such a Change

9.1 Language Change or Language Deterioration?

While "historically, asynchronous communication, like letter writing, has tended to be more formal and less off-the-cuff exchanges" (Negroponte 1995: 167), today's pendulum swings in the other direction and asynchronous communication has become more informal, especially as far as computer-mediated communication is concerned. Baron (2000: 188 f.) explains that "writing has […] become more speech-like, mirroring informal conversation. Increasingly, we see writing as an interchangeable alternative to speech rather than a distinct medium with its own functions and conventions". Petrie (1999: 3) even speaks of a "renaissance in written communication". Yet, there has been an ongoing discussion how far this renaissance is actually accompanied by language deterioration. In an interview at the University of Vienna, Schlobinski (2002: 2) states that the writing of SMS is not necessarily incorrect, but characterised by the use of new forms, whereas the *Sunday Herald* titles "Teachers call for urgent action as pupils write essays in text-speak" (Johnston 2003). Apparently, text-speak has broken its boundaries, and so the Scottish Qualification Authority finds that "text-message language was inappropriately used in the writing section of last year's English exam", while a Scottish English teacher says that "she was shocked to read an essay by a second-year pupil written in text message shorthand" (Johnston 2003). Meanwhile, the aforementioned essay has become the example par excellence when debating about language deterioration:

> My smmr hols wr CWOT. B4, we usd 2 go 2 NY 2C my bro, his GF & thr 3 :-@ kds FTF. ILNY, its gr8. Bt my Ps wr so {:-/ BC o 9/11 tht they dcdd 2 stay in SCO & spnd 2wks up N. Up N, WUCIWUG -- 0. I ws vvv brd in MON. 0 bt baas & ^^^^^. AAR8, my Ps wr :-) -- they sd ICBW, & tht they wr ha-p 4 the pc&qt…IDTS!! I

wntd 2 go hm ASAP, 2C my M8s again. 2day, I cam bk 2 skool. I feel v O:-) BC I hv dn all my hm wrk. Now its BAU … (*Spiegel* 2003)[1]

The attitudes towards text-speak diverge. While Judith Gillespie, a development manager at the Scottish Parent Teacher Council, maintains that "there must be rigorous efforts from all quarters of the education system to stamp out the use of texting as a form of written language", Sheila Hughes, a senior lecturer of language education at Strathclyde University, believes that "if teenagers express themselves well with the medium it should be embraced", explaining that "texting is a great opportunity for teaching to explore a new use of language. Pupils are thinking about vowels and punctuation even if they are dropping them. You can't stop language changing so let's just exploit it – pupils have power and control of that language and it has a purpose we have to value" (Johnston 2003). Similarly, Beutner (2002: 91) explains that language has always been changing and always will; there can be no question of language deterioration by the influence of the new media. In tune with Beutner, Moraldo (2004b: 267) finds that despite the great worries from many sides, there has not been any language deterioration as yet, while Thurlow (2003) has a very pragmatic and progressive view towards text-speak: he asserts that "[t]he notion of standardness in written language is itself a convention and always an abstraction from spoken language (see Cameron, 1995); in this sense […] many of the typographic practices of text-messaging offer more 'correct', more 'authentic' representations of speech". Finally, Schlobinski (2002: 3) concludes that it would not be sensible to keep young people from SMS communication because people who do not acquire competence in the use of this medium will sooner or later belong to the losers of the media society.

1 The translation is: My summer holidays were a complete waste of time. Before, we used to go to New York to see my brother, his girlfriend and their three screaming kids face to face. I love New York, it's great. But my parents were so worried because of the 11 September that they decided to stay in Scotland and spend two weeks up north. Up north, what you see is what you get – nothing. I was very very very bored in the middle of nowhere. Nothing but sheep and mountains. At any rate, my parents were happy – they said it could be worse and that they were happy for the peace and quiet… Idiots!! I wanted to go home as soon as possible, to see my mates again. Today, I came back to school. I feel very saintly because I have done all my homework. Now it's business as usual …

When evaluating text-speak, the problem is also one of age and generation. Some linguists assert that "the abbreviated language creates a teen-only zone that adults cannot easily enter" (*ONE News* 2003). Accordingly, texting teenagers have lately been described as 'generation txt'. Ling and Yttri (2002: 162) come to a similar conclusion, stating that the use of text-speak also marks group boundaries; it allows the users "to establish an identity separate from that of parents, teachers and other adults". In this context, text-speak can be regarded as an in-group code and as such serves a certain identification function. Yet, what has to be noted is that this new language is not completely disorderly and thus not unintelligible. There are recurring patterns according to which the language is condensed, such as the dropping of vowels in consonant clusters or the use of letter and number homophones. In fact, numerous creative new forms have developed out of CMC, mostly for economy and efficiency, and if one is familiar with text-speak, the messages can be read without difficulty.

Even though occasionally feared, it is unlikely for various reasons that other more formal genres will ever make use of text-speak. First, this reduced writing style is not suitable for the swiftness of handwriting or typing (as opposed to the laborious keying in of words on a mobile phone keypad) and second, text-speak is the language of immediacy and unplanned discourse, whereas formal writing is planned and not composed under real-time constraints. Moreover, the above cited student essay does not correspond to the reality of text-speak, as can be deduced from this study. Text messages are, on the whole, less cryptic than commonly claimed. Hence, it can be concluded that language as a whole is not deteriorating as long as text-speak is used in its specific environment. Rather, the use of language in CMC can be seen "as evidence of the uniquely human ability to adapt language according to the interdependent variables of a given situation (context, activity, relation to communicators), level of synchronity [and] means of expression (writing vs. speaking, production and perception conditions of the means of expression)" (Segerstad 2002: 29). As soon as it intrudes into other genres, however, and affects text types such as essays or any other formal writing, it would be a sign of language deterioration insofar as it would displace the cultivated language of planned discourse. But within their medium "language uses in CMC are variants of use, just like any other language use" (Segerstad 2002: 29). It is, of course, possible that features common to CMC will eventually affect formal and planned writing as well, especially with the increasing number of people relying on electronic discourse. Whether or not CMC will eventually influence other genres, however, remains to be seen and cannot be predicted at this stage.

9.2 New Trends – Old Features

Many features of CMC and particularly text-speak are not actually new. Bergs and Kesseler (2003: 82) come to the conclusion that "SMS have surprisingly similar traits to common letter-writing in the nineteenth century. People use it mostly for giving short reports, making appointments, sending regards, but also for declaring their feelings" (cf. also Siever 2005). Linguistically, there is a significant dominance of lexical reductions over syntactic ones in SMS (in emails, lexical and syntactic reductions are more balanced), which is not only due to economy, but also constitutes a matter of simplification to approximate the sound system to orthography. Lexical reductions may lead to spelling variation so that the word 'tomorrow', for instance, can be spelt in at least nine different ways. This spelling variability has led to what Baron (2000: 104) would call a "laissez-faire" attitude towards orthography, which leads us back to the time before the end of the 17th century, when spelling was not yet standardised (Barber 1993: 201). Hence, whereas there are at least nine different ways of spelling the word 'tomorrow' in CMC, namely *tomorrow*, *2m*, *2moro*, *2morrow*, *2moz*, *2mz*, *tomm*, *tomoz*, and *tmw*, in Middle English there were four different ways, namely *to-morwen*, *tomorwe*, *tomoreȝe*, and *tomorn* (Mayhew and Skeat 1978: 234).

A feature which dates back to fifth or fourth century BC Greece is the lack of spaces between words (Baron 2000: 169). This feature now recurs in CMC: in emails, but more commonly in text messages, two or more words can be strung together and written without spaces in between, as Table 9-1 indicates.

Table 9-1: Words Written Without Spaces in Between

Number of words	per 1000 words in emails	per 1000 words in single SMS	per 1000 words in linked SMS
2	1.13	4.54	2.56
3	0.02	2.55	1.39
4	0.00	0.33	0.46
5	0.00	0.11	0.23
8	0.00	0.06	0.00
total	1.16	7.59	4.65

Examples (269) to (271) show that it is often words with number homophones that are combined, so that a visual division between the individual items nevertheless exists.

(269) Thank u so much *4coming2meet* me (My emphasis)
(270) what u *going2wear2nite*? (My emphasis)
(271) I know the best thing *is2 talk2hm* but i dont want *2get* hurt and i just want an opinion on it. (My emphasis)

In contrast to ancient Greece, however, it is only words and not sentences that follow one another without spaces in between, i.e. words across sentence boundaries are not spacelessly attached to each other even though, similar to then, punctuation marks may be omitted. Yet, it may happen that spaces are not put after punctuation marks, as can be seen in examples (272) and (273). This is well worth noticing, but Table 9-1 above does not count these instances because the words are separated from each other by some form of punctuation.

(272) Aww thnku [name] ur da *bomb!&thnkx* so much4da card absolutli *luvli!luv u!xxx* (My emphasis)
(273) Sure,no *worry!What's* happening with your stay *arrangements?You* cant get through to *sarah?You* could always stay *here,tom* is down so would be menage a trois! (My emphasis)[2]

2 For reasons of clearness, names are not substituted by the common placeholder [name] in this text message. Yet, to still guarantee anonymity, pseudonyms are applied.

Moraldo (2004a/b) points out that the readability of strings of words written without spaces between them is mostly guaranteed by majuscules marking the word boundaries (Moraldo 2004a: 295 and Moraldo 2004b: 256). This, however, appears to be a German[3] rather than an English feature: whereas 77% of all the messages which contain spaceless sequences of strings of words are indeed marked by majuscules in the German corpus, there is no such example found in the English corpora. One of the reasons for this finding is that German messages hardly contain number homophones. The only number homophones which may appear in German text messages are *8* and *1*. Following Moraldo (2004a: 293), the cardinal number *8* is used in the combination *gN8* (meaning 'gute Nacht', i.e. 'good night'[4]), while the number homophone *1* may sometimes be used as indefinite article. Yet, they seem to be rare and so, there is only one number homophone in the German corpus, this being the one in example (274):

(274) Hey babe. gehst mit schlafen? wünsch dir ne gute nacht. soll i dich morgen wecken? dickes *guten8bussi*, hdl ps.ich träum von dir :) (My emphasis)[5]

Hence while it is number homophones which are used to mark word boundaries in spaceless strings of words in English messages, it is majuscules in German messages, as examples (275) and (276) demonstrate.

(275) DuSchmeichler![Name]IchVersprecheDirBaldMehrZeitZuHabenWirKönnen GerneMorgenTelef.!JetztSitzeIchNochüberMeinemAbschluss*GrüßeUnd GuteNacht*C[6]

---

3 The same feature also applies to Italian messages (cf. Moraldo 2004a: 295).

4 Moraldo (2004a: 293) explains that *gN8* seems to be borrowed from English, meaning 'good night' where this alpha-numeric combination allegedly exists, too. Yet, there is no instance of such a combination in the analysed corpora. Also, it would linguistically not be appropriate to use the number *8* to spell *night* for the phonological value is not the same: while *8* is pronounced [eɪt], *night* is pronounced [naɪt]. In German, however, *8* [ʔaxt] corresponds to *N-acht* [n-axt]. Hence *gN8* rather seems to be a German version of spelling 'gute Nacht' than an English version of spelling 'good night'.

5 Hey babe. are you coming to bed, too? i wish you a good night. shall i wake you up tomorrow? big good night kiss, i love you ps.i'll be dreaming of you :)

6 YouFlatterer![Name]IPromiseToHaveMoreTimeSoonWeCertainlyCanPhone Tomorrow!ImStillSittingOverMyFinalExamPreparations*GreetingsAnd GoodNight*C

(276) HiMeinSchatzWiegehtsDir?WarHeuteNichtAufDerPisteMirGehtsNicht GutHoffeDasIchMorgenWiederFahrenKannTrinkenGradMariandlDasIstSo NeArtSekt.RedenGradÜberDichIchVermissDichFreuMichAufDichHolstDu MichAbWillDannSchlafenUndAbendsWeg.LiebDich!MeldeMichNoch. VonWoSchreibstDuDieSMS?WasMachstDuDenGanzenTag?[7]

9.3 Possible New and Generally Accepted Conventions

Computer-mediated communication is unplanned discourse and therefore spelling mistakes and erroneous language are largely tolerated. They do not reveal any gap in the sender's orthographic knowledge, but are excused by the swiftness with which messages within CMC are usually composed, there being little or no editing at all. To judge the quality of a given message is difficult in any case: there are no relevant rules available for the language used in computer-mediated communication just as there are no rules for unplanned discourse. Hence, traditional style manuals often seem inadequate for the composition of emails and text messages, and those which are made for electronically transmitted messages "differ over the question of just how much editing should take place: on the one hand, they are anxious to maintain their belief in the medium's informality; on the other hand, they are driven by the awareness that, the more idiosyncratic behaviour departs from the norms of standard English, the greater the likelihood of unintelligibility" (Crystal 2001: 111).

The evolution of dictionaries is one possibility to legitimize the unplanned discourse of email and text messages in a more official manner and make it more widely accepted within the field of computer-mediated communication while still aiming at a certain orthographic

7 HiMyDarlingHowAreYou?IWasn'tOnThePisteTodayIDon'tFeelWellHopeThatI CanGoAgainTomorrowDrinkingMariandlWhichIsAKindOfSparklingWine.TalkingAboutYouIMissYouLookingForwardToSeeingYouWillYouPickMeUpIWantTo SleepThenAndGoOutInTheEvening.LoveYou!IWillTalkToYouSoon.WhereDo YouWriteThisSMSFrom?WhatHaveYouBeenDoingTheWholeDay?

consistency and legibility. *Lingo2Word* is an example of an online text messaging dictionary which translates text messages from 'lingo' into plain English and vice versa. It also offers various lists of emoticons, texting vocabulary and acronyms. Such online dictionaries, however, are still in the process of development and the problems are manifold. For instance, *Lingo2Word* lacks a considerable number of words that are not in the translator: it uses capitals where they would never be employed (e.g. *sA* for 'say'), and its translations do not always correspond to practice or are inaccurate if not wrong (e.g. *thort* for 'thought'). Also, it is rather improbable that an emoticon is used instead of a nickname, as emoticons are usually employed to express emotion and do not constitute a means of lexical reduction (e.g. ~:o for 'babe'). Quotations (277) and (278) provide contrasting examples of the actual wording employed in the given text messages and the corresponding text message lingo translation.

(277)	SMS:	No im not cummin2twn! ive got stuff2do-media stuff! cu 2moz!
	Text Lingo:	No im nt cmng 2 twn! Ive got stuf 2 du – media stuf! CU 2moz!
(278)	SMS:	Hey babe, at work, Jst thot id say hi, hope works not 2 bad
	Text Lingo:	ey ~:o, @ wrk, jst thort Id sA hi, hope worx N2B

Yet, despite its inaccurate translations, *Lingo2Word* can nevertheless be regarded as a step towards a more general acceptance of text-speak, and the same is true of the online *Texting Dictionary* launched by HarperCollins in 2004, where one finds a relatively comprehensive list of texting abbreviations. Some of these abbreviations already feature in the new edition of the Oxford English dictionary, while small-format text message dictionaries have been available since 2000, when the *ltle bk of txt msgs* (Mander 2000) was published. It contains over 1000 abbreviations and emoticons as well as their meanings. These dictionaries contribute to a certain standardisation of text-speak. According to Haugen (1996: 109), this standardisation process involves four steps, namely

1. selection of norm,
2. codification of form,
3. elaboration of function,
4. acceptance by the community.

In other words, the lexical items to be included in the dictionary must first be selected, i.e. it must be agreed on "the selection of some kind of a model from which the norm can be derived" (Haugen 1996: 109). Then, the "minimal variation in form" (codification) and "maximal variation in function" (elaboration) have to be defined and finally, if a standard variation "is not to be dismissed as dead, [it] must have a body of users" (Haugen 1996: 109) and become generally accepted. Once the language of text-speak is officially acknowledged and included in dictionaries, it will certainly gain greater acceptance also by those people who have so far disapproved of text-speak. Yet this is not to say that text-speak will lose its ludic mode; rather, individual spelling and language-play will persist to a certain extent because texting has become a grammar-/spelling-rule-free writing-zone which is enjoyed and experienced to the full.

9.4 Does Text-Speak Foster a New Form of Illiteracy?

It may be argued that spelling variability and the laissez-faire attitude towards language in computer-mediated communication lead to a new form of illiteracy. While composing text messages, users do not aim at a carefully chosen language, for it is no longer style that matters but information, and the credo is *as much information as possible with a minimum of signs*. A condensed language is the result. Abbreviations, however, are not a new phenomenon:

> Often thought to be an exclusively modern habit, the fashion for abbreviations can be traced back over 150 years. In 1839, a writer in the New York *Evening Tatler* comments on what he calls "the initial language [...] a species of spoken shorthand, which is getting into very general use among loafers and gentlemen of the fancy, besides Editors, to whom it saves much trouble in writing [...]". (Crystal 1995: 120)

Initial language is also known in computer-communicated conversation where words or syllables are abbreviated in such a way that only the first letter is written. Examples (279) to (282) contain instances of initialisms and acronyms: *btw* and *asap* refer to the stereotyped phrase 'by the way' and 'as soon as possible', whereas the letter *b* in quotation (281) means 'be', and *TB* in the last example (282) is short for 'text back'.

(279) I have no reception at this customers *btw* and should fin soon (My emphasis)
(280) i ill spk 2hm *asap*&let u know! (My emphasis)
(281) I will *b* in wild boar at 8.30pm. (My emphasis)
(282) Hi [Name] its [Name] is my New Moby No Wot ru up2? Glad xams rover *TB* luv [Name] (My emphasis)

Crystal (1995: 120) explains that "the fashionable use of abbreviations – a kind of society slang – comes and goes in waves, though it is never totally absent". But with the advent of SMS, abbreviations seem to have reached a new climax with a new and so far unknown potential: there is a ludic moment inherent to the composition of text messages which enables users to play and experiment with the language. The aim is always the same: to save time and space. Playing and experimenting with language, however, does not foster illiteracy, for people still follow rules that are based on shared lexical and grammatical knowledge, as has been pointed out in Section 3.5.2.3, Non-Conventional Spelling (cf. also Schlobinski 2001a). There is certain regularity, but the problem is that these rules have not officially been established nor have they acknowledged acceptance and codification. Nevertheless, there must be some underlying conventions for otherwise, text-speak would be totally unintelligible. Yet, these conventions are significantly more flexible and versatile than those of planned discourse.

In short, computer-mediated communication is a new type of discourse. It is written (near-)synchronous discourse that has been developing its own language – a language that is suitable for the immediacy of (near-)real-time written communication. It constitutes a new register with new purposes that is equally justified as any other text type such as the writing of postcards or memoranda. Hence, it can be assumed that text-speak does not foster illiteracy and, as Bergs (2003: 8 f.) correctly states, "neither has email […] led to significant changes in any linguistic system around the world nor has text messaging led to language structures or uses that have not been there at one point in time and have been recorded for other means of communication, such as telegrams or postcards".

9.5 What Influences Language Change?

What would happen to email messages if they were – as text messages – retrievable from a mobile medium and no longer stationary? Would their language change and become more condensed or would it remain the same? In other words, is it the medium that influences language change? As pointed out in Section 1.3, Methodology: Data Collection and Corpora, the average text message does not use up all characters that are available, but only about half of them. Even if the words were not abbreviated, the messages would not necessarily add up to 160 characters, so that the space restriction is not the only reason for the reduced writing style of text-speak. Ergo: the major factor of language change in computer-mediated communication cannot be the medium itself, i.e. the computer or mobile phone; instead, there must be some other, weightier factor, this being immediacy. Accordingly, Maynor (1994: 53) explains that "many of the differences between speech and writing are a result of the immediacy of speech", and Segerstad (2002: 19–21) establishes the following maxim: "the more synchronous the mode, the more features it shares with spoken face-to-face interaction", i.e. "the more time pressure is exercised on the speed of the typing and transmission". If one can expect the recipient to be reachable everywhere he goes, one sends messages in view of this immediacy and expects a correspondingly immediate answer which does not allow the sender to carefully think about the chosen language and compose novelistic messages. Similarly, as has already been shown above (cf. Chapter 4, Single Text Messages versus Linked Text Messages), Thurlow (2002) argues that "the need for both brevity and speed appears to be motivated less by technological constraints, but rather by discursive demands such as ease of turn-taking and fluidity of social interaction". Segerstad (2002: 25) further points out that it is also the "situational parameters such as relationship between communicators, and goal of interaction, together with parameters associated with means of expression, such as production and perception conditions" that influence the language of electronic messages. Hence, what might change once email services become mobile is the chronemic, i.e. the time planning of communication, plus the length of emails, which might become significantly shorter. Whether or not the style of writing will change remains to be seen and also depends on the kind of mobile keypad/-board which will certainly encourage a change in style if its handling is as laborious and time-consuming as that of today's mobile phones.

As for the speech-like nature of computer-mediated communication, it has to be noted that the shift towards conceptual orality is in line with tendencies towards orality starting long before the advent of email and text messages: there has been a general trend towards orality since the late 19th century, which computer-mediated communication may have accelerated. Thus Baron (1998b: 52) recognises that

> in the nineteenth century, 'composition' was typically oral, though modelled on written standards. In the twentieth century, writing is the pedagogical norm, though increasingly modelled on speech. Computer technology has influenced the teaching of writing, but generally in the directions that composition theory and broader educational philosophy were already leading. To the extent that composition is increasingly done on-line, it seems likely that the spoken-language properties of email will reinforce the increasingly speech-like character of writing that school and college pedagogy [...] has been fostering since the late nineteenth century.

9.6 Anglicisms in German and Swiss-German Emails and Text Messages

Siever (2005: 146 f.) explains that while studies on letter communication from the 19th century have already shown features which can be considered typical of CMC (cf. also Bergs and Kesseler 2003), it is mainly the use of Anglicisms which has become more frequent. In fact, the English language has a special significance: it can be regarded as the *lingua franca* in our world and the first universal language in the history of mankind (Hoberg 2000: 303). Its influence has considerably increased since the end of the Second World War (cf. Hoberg 2000, Watts 2004, and Wilss 2003), but at the same time, it has become highly controversial and has been frequently debated ever since. Within the Germanic domain, Anglicisms are usually regarded negative, as the simple title of the following book, which comprises a selection of essays on the topic, implicates: *Denglisch, nein danke!* (Zabel 2003). The Society of the German Language, known as Verein Deutsche Sprache, even demands governmental measures against Anglicisms (Stickel 2004: 19). The motives for the dismissive stance towards Anglicisms in the German language are various. On the basis of approximately 330 letters on the cause of this negative attitude, Hoberg (2004: 312–314) lists four main reasons:

1. redundancy,
2. incomprehensibility,
3. a means to show off and impress,
4. identity problems.

First of all, people regard Anglicisms as superfluous: the newer they are, the more they distrust them. Hoberg points out that one of the Anglicisms that is frequently held to be redundant is the word *Kids*. Yet, he explains that *Kids* is by no means identical with *Kinder*[8]. No one who wants to know how many children his vis-à-vis has would ask: "Wie

8 children

viele *Kids* haben Sie?"[9] *Kids*, Hoberg explains, does not appear in the singular in German; it is a collective term which is most often ironically used by people in their middle age for today's youth, chiefly for the children of the McDonald's culture, and displaces older terms such as *Teens* or *Teenager*. A second reason why people disapprove of Anglicisms is that they claim that many people do not have good command of English so that communication is unnecessarily complicated. Third, they perceive the use of Anglicisms simply as a means to show off or impress and last, they believe that the Germans have problems with their (national) identity: they have low self-esteem and thus abandon the German language far too easily (Hoberg 2000: 313 f.).

Nevertheless, the number and use of Anglicisms has increased despite the vast criticism. Especially since the introduction of the mobile phone, Anglicisms seem to have become more wide-spread: a comparison of German and Swiss-German emails and SMS clearly shows that Anglicisms are used twice as much in text messages as in emails. In fact, it is highly interesting to observe how quickly German and Swiss-German texters code-switch to English vocabulary. Examples (283) and (284) illustrate this finding:

(283) AKKU GRAD LEER ESS SCHNELL WAS U *CALL* IN 20 MIN CA *SORRY HOPE OK* (My emphasis) [10]

(284) Hallo [name]! danke für dini iladig! ich bin nächscht wuche 1nacht in dütschland, weiss abr nonig wänn. susch chumi gärn! *what about blue monkey*? liebs grüessli (My emphasis) [11]

In Table 9-2, the most frequently used Anglicisms that appear in the analysed German and Swiss-German corpora are listed and classified according to their frequency.

9 How many kids do you have?

10 THE BATTERY IS FLAT I QUICKLY EAT SOMETHING AND CALL IN ABOUT 20 MINS SORRY HOPE OK

11 Hello [name]! thank you for your invitation! i'll be in germany for one night next week, but i do not know when. otherwise, i'd love to come! what about blue monkey? love

Table 9-2: Anglicisms in (Swiss-)German Emails and Text Messages

	per 1000 words				
	in GER		in SUI		
Anglicisms	emails	SMS	emails	SMS	%
oke	0.00	0.00	0.00	0.49	1%
bye	0.07	0.16	0.09	0.21	1%
internet	0.36	0.00	0.19	0.07	1%
kiss/es	0.00	0.08	0.00	0.56	1%
cu ('see you')	0.00	0.08	0.09	0.56	1%
okay	0.03	0.48	0.14	0.14	1%
weekend	0.00	0.24	0.09	0.63	1%
mb ('mail back')	0.00	1.04	0.00	0.00	1%
easy	0.00	0.00	0.38	1.04	2%
party	0.73	0.24	0.38	0.49	2%
handy	0.63	1.28	0.19	0.07	3%
cool	0.30	0.48	0.24	1.53	3%
sorry	0.23	1.12	0.66	1.74	4%
e/mail	2.15	1.20	1.42	0.42	6%
ok	0.36	2.95	0.09	2.79	7%
hi	0.73	6.54	0.24	0.77	10%
hey	0.60	5.67	1.09	6.69	16%
others	8.31	5.51	8.46	12.89	41%
total	14.50	27.05	13.76	31.07	100%

Dürscheid (2000: 243 f.) also points out that often English diction appears where the words do not even contain any English components and thereby refers to compounds which are usually written as one word, but all of a sudden appear as two words in German texts. This corresponds to English spelling where nouns can be strung together in order to build a complex expression. Such two-word compounds, Dürscheid explains, are often found in texts of the software industry. Accordingly, this feature hardly occurs in the analysed corpora and is not further analysed. One of the very few two-word compounds, which consists of purely German components but is written as two words (although with a hyphen), is the compound *sozial-indikatore* (meaning 'social indicators') in example (285).

(285) alloä! mer hend di falsche tascheräächner! nanana, hesch gspickt! hehe! die 12 *sozial-indikatore* ;) du, hemmer im rw öbis i dem blaue buech? hdl! greez! (My emphasis)[12]

English diction in words without any English components, however, may also affect the spelling of the postalveolar fricative [ʃ]. In Swiss-German emails and text messages, the postalveolar fricative [ʃ] of /sch/ in words like *gshänkli* (meaning 'present'), *wünsh* (meaning 'wish'), *shwätze* (meaning 'talk'), or *shatz* (meaning 'darling') happens to be spelt /sh/, thus corresponding to English orthography where /sh/ is pronounced [ʃ]. This feature does not appear in any of the German corpora and only occurs once in the Swiss-German email corpus. In the Swiss-German SMS corpus, in contrast, it is counted 1.04 times per 1000 words. Quotations (286) to (288) provide examples of this feature:

(286) [Name]..Hüt simmr 7h wartend im unispital xi mit muulchorb [mundschutz] &min [name] hät ez *warsh* [wahrschiinlich] doch tb. Uf morn *ish* är *sletshtmal* dihei,morn morge nämeds än mr wäg für mind 2wuche:-(scheisse. Ha di nur wele informiere..Xemr eus doch gli,dänn chömer usgibig übr ales *shwätze*. Guet nacht,dear sister! Bis bald,[Name]. (My emphasis/comment)[13]

(287) Sorry [name]! Han am suntig es fotoshooting füres heftli! Isch sehr churzfristig cho…ahhh!Jetzt chan i nid cho, das gehbt es mega gstürm! Zug *ish* wit. Lg [Name] (My emphasis)[14]

---

12 hello! we've got the wrong calculators! nanana, you have had a crib! hehe! the 12 social indicators ;)by the way, do we have any homework in accountancy in the blue book? i love you! greez!

13 [Name]..Today, we have been waiting for 7 hours at the university hospital with muzzle [mask] and my [name] seems to have tuberculosis still. Tonight, he will be a last time at home and tomorrow, they'll take him away from me for at least 2weeks:-(shit. I've just wanted to let you know..Lets meet soon to talk about everything more thoroughly. Goodnight,dear sister! See you soon,[Name].

14 Sorry [name]! I'm having a photo shooting for a magazine on Sunday! I've got it on very short notice…ahhh!Now, I cannot come, that would be too much trouble! Zug is far away. Love [Name]

(288) ciao [name].so, *häsh* e *shöns* WE ka?wills doch hoffä.ich wünsch dir ganzen guetä start idi noi wuche.pfus guet.bis baud,liebs grüessli [Name] (My emphasis)[15]

Examining Table 9-2, it can be concluded that the result is the same for the German as well as the Swiss-German language: both languages show a similar use of Anglicisms when it comes to the composition of emails and text messages. Overall, the interjections *hey* and *hi* are most frequently used, followed by *ok*, *e/mail*, and *sorry*. There is, however, an interesting difference in the use of certain Anglicisms: some of them appear in German messages only while others merely occur in Swiss-German ones. A common German Anglicism is the abbreviation *mb*. It stands for 'mail back' and is usually added at the end of text messages as illustrated in examples (289) and (290).

(289) Bist du nächstes we daheim? Würd voll gern mal deine „neue" wohnung sehen! Wär echt mal wieder cool zu ratschen mit dir. *Mb* (My emphasis)[16]

(290) Ja bin nächstes we freitag abend und sonntags daheim.meld dich halt dann nochmal wann du zeit hast,ja?*mb* (My emphasis)[17]

Interestingly, none of the English text messages contains this abbreviation. In English, people use *tb* (meaning 'text back') instead, as examples (291) and (292) show.

(291) Hey. This place is a bit boring now. Stil better thn college. Tell peeps i say hi. *Tb* (My emphasis)

(292) Yo!Wat is happenin bro! U cool? I herd u crashed you're car,wat happened!!*Tb* (My emphasis)

The *mb*-feature at the end of German text messages seems to be an exclusive German Anglicism, as it is used in a way Anglophone people would never do nor does it exist in the Swiss-German corpora. Yet, there

15 ciao [name].well, have you had a nice weekend?i do hope so.i wish you a very good start into the new week.sleep well.see you soon,love [Name]

16 Are you at home next weekend? I would love to see your "new" home! It would be cool to have a chat again. Tb

17 Yes, I'm at home on friday evening and sunday next weekend.get in touch with me again when you have time,ok?tb

are also Anglicisms which are exclusive to Swiss-German emails and text messages, such as the following two lexical items: *easy* and *oke*. The word *easy* is comparably frequent in the Swiss-German corpora, especially in text messages where it appears 1.04 times per 1000 words. It can be used in various contexts as an adjective as in quotation (293), as an adverb as in quotation (294), or as an interjection as in quotation (295).

(293) Heey! Ja voll!und bi dir? Was isch scheisse gsi?! ooh:(jaja han au es *easy* weekend gha!Is extra? Hmm,du gasch?Lg (My emphasis)[18]

(294) Ich hängsi gad mim VWL-Buäch im Bett. Gad mär *easy* und dir? Ne, kuul ä mäntig wird hüt… (My emphasis)[19]

(295) *Easy*! Au dir en schöne abig und bis bald! (My emphasis)[20]

In contrast to the word *easy*, *oke* is less frequent but still significant. It does not appear within the email corpus, but is counted 0.49 times per 1000 words in the text message corpus. The word *oke* is a Helvetic way of pseudo-phonetically spelling *okay* and can thus be regarded as a Helvetisation of the English word *okay*. Examples (296) and (297) illustrate its use.

(296) Müseli mir sind am 20 vor 10i in pfäffike!? *Oke*? (My emphasis)[21]

(297) ähm ja chunsch du mi go holä?! wär das *oke*?! Ja so gäg dia 9i, oder wa meinsch?! wo, ja luege mer spontan,oder?! (My emphasis)[22]

There are four major forms of Anglicisms. The simplest and most frequent ones are the direct borrowings, i.e. those Anglicisms which are directly adopted from the English language without being modified, so-called genuine Anglicisms. These may be simple elements as the adverb *again* in example (298), compounds as *mailbox* in example (299),

18 Heey! Yes of course!and you? What was so bad?! ooh:(yes yes I also had an easy weekend!To the extra? Hmm,are you going?Love

19 I'm chilling in my bed with my economics book. I'm fine and you? No, it will be a cool monday today…

20 Easy! Have a nice evening yourself and see you later!

21 Honey we are in pfäffikon at 20 to 10!? Ok?

22 ehm yes will you pick me up?! would that be ok?! Well at 9ish or what do you think?! where, well, let's decide spontaneously,ok?!

phraseologisms as *See ya later, alligator* in example (300), or initialisms as *btw* (meaning 'by the way') in example (301).

(298) SCHICK MAIL *AGAIN*,MÜSST NOCH DRIN HABEN.WARUM MACHT DER DAS?DA WOLLT DER MICH NOCH NACH L BRINGN,HAT GREDET MIT DIR.ES TUT MIR SO LEID,DU HAST SO LANG GKÄMPFT.IN LON.KENN NUR ADRESS FÜR SOMMR.HÄTT GERN GHOLFN.KANN I DICH MAL CALLN?KOPF HOCH DENK A TI GANZ DOLL!DICKE UMARMUNG (My emphasis)[23]

(299) Vielleicht kannste dich ja ma aufraffen und wat schreiben. Musste aber nich. Ich schreib dir auch wenn du nich schreibst. Hehe. Musst sagen wenn ich deine *mailbox* zumülle und dich nerve. (My emphasis)[24]

(300) Wenn Du Lust hast können wir uns ja mal treffen. Vielleicht erst mal an der Uni, ich studiere auch hier Englisch und MusikWissenschaft. *See ya later, alligator*!? Gruß [Name]!!! (My emphasis)[25]

(301) Hi darling.Do schint grad d'sonne,aber ohni dech ergendwie nor halb so hell. Vermesse dech.Gods der besser?*Btw*,en guete (My emphasis)[26]

Table 9-3 illustrates that, on average, genuine Anglicisms appear 18.03 times per 1000 words. They are considerably more frequent than any other category: 84% of all Anglicisms are genuine ones. This result is on par with Busse (2004: 83), who points out that in most cases, Anglicisms are borrowed directly. The verb *calln* in example (298), however, also indicates that Anglicisms can be added a flexion. Other such examples are shown in quotations (302) to (305): *chillen*, *beachä, sleepe*, and *geloudet.* These are Germanised Anglicisms, i.e. words which are not only borrowed, but also adapted so as to fit the German grammar rules.

---

23 I'LL SEND THE MAIL AGAIN,SHOULD STILL HAVE IT.WHY DOES HE DO THAT?HE EVEN WANTED TO TAKE ME TO L,TALKED TO YOU.I'M SO SORRY,YOU'VE BEEN STRUGGLING SO LONG.IN LONDON.ONLY KNOW AN ADDRESS FOR THE SUMMER.WOULD HAVE LOVED TO HELP YOU.CAN I CALL YOU SOMETIME?CHIN UP I'M THINKING OF YOU!BIG HUG

24 You might pull yourself together for once and write something. However, you don't have to. I'll write you even if you don't write. Hehe. You must tell if I glut your mailbox and get on your nerves.

25 If you fancy we can meet sometime. Maybe at university at first, I study English and musicology here, too. See ya later, alligator!? Love [Name]!!!

26 Hi darling.The sun is shining here,but it is only shining half as bright without you.I miss you.Do you feel any better?Btw,enjoy your meal

Table 9-3: A Classification of Anglicisms

	per 1000 words				
	in GER		in SUI		
Anglicisms	emails	SMS	emails	SMS	%
pseudo-Anglicisms	0.66	1.36	0.28	0.07	3%
hybrid Anglicisms	2.25	0.56	1.13	1.11	6%
Germanised Anglicisms	1.19	0.48	1.80	3.34	8%
genuine Anglicisms	10.40	24.65	10.54	26.54	84%

(302) Hallo!Du ich bin so platt heute& froh, wenn ich bissel *chillen* kann:-)Solltet ihr euch heut abend treffen, kanns dus mir sagen,viell. komm ich nach! (My emphasis)[27]

(303) aber jetzt chunnt ja hoffentlich en schönä früehlig ;-) dänn chasch du go *beachä* und ich go zueluegä... *lol* (My emphasis)[28]

(304) Huidui... Jetzt beni grad vo mine eltere onde ufe cho... D'[name] esch au no do! Ui, gesch es mega loschtig gsi im reschti, verzell der denn! Wie gohts der? Ech gang jetzt hurtig go *sleepe* ;-) müntschi (My emphasis)[29]

(305) Rate mal, was ich hier in meinem Rucksack habe! Musste 2 Läden durchforsten danach, Seth ist der Coolste! Hab auch schon die 2. Staffel *geloudet.* Wird cool. (My emphasis)[30]

Another category of Germanised Anglicisms constitute those words which have a CONS-*y* ending. This is a *y*-ending following a consonant as in *party*, *baby*, or *lady*. According to the English grammar rule, nouns ending on /y/ form their plural on /ies/ after a consonant. This rule, however, does not apply to Anglicisms in German and Swiss-German emails and text messages. Accordingly, the words *baby* and *party* in examples

27 Hello!I'm so knackered today& and glad if I can chill a bit:-)Should you meet each other tonight, you can tell me,I may come later!

28 but now a beautiful spring is coming, hopefully ;-) then you can go to the beach and I can come and watch... *lol*

29 Hey you... I've just come back from my parents... [name] is also here! Oh, it was really funny yesterday in the restaurant, I'll tell you! How are you? I'll quickly go to bed now ;-) kiss

30 guess what I've got in my backpack! I've had to plough through two shops for it, Seth is the coolest one! I have already loaded the second season. Will be cool.

(306) and (307) contain a CONS-*ys* plural. This, in fact, corresponds to the new German orthography which no longer allows the CONS-*ies* ending, but postulates the CONS-*ys* ending instead.

(306) diese woche hab ich mich erkaeltet, das klima hier ist sehr abwechslungsreich, deswegen hatte ich ziemlich husten. ich ging darum heute nicht arbeiten, um die *babys* nicht anzustecken. (My emphasis)[31]

(307) copol monday im palais isch eini vo dä absolut geilstä *partys*. dettä muäsch ga. ich gangä dänn au wieder wänni nümmä in grüän bin. (My emphasis)[32]

Moreover, Germanised Anglicisms may show highly idiosyncratic spellings. The English verb 'to work', for instance, may appear as *worke* (308) or as *wörkn* (309). Or else the word 'cool' may be spelt *kuul* (310) or simply *ql* (311), while 'easy' is pseudo-phonetically spelt *isi* (312) and so is the verb 'to load' which appears as *geloudet* in example (305) above.

(308) Jetzt goht es fuuls manndeli ganz fescht go pfüsele! ;-) be höt weder voll id arbetswält iitaucht!! Muess öbrigens die next 2 sa ned go *worke*… Schmötzi (My emphasis)[33]

(309) Oh [Name]…bi do gad chli we verruckt am *wörkn*…meld mi gad gli…e Oringles faucht au nu…grummel (My emphasis)[34]

(310) wet frogä ob er no en pröfig vodä [name] hend,thema esch wärmelere!wär *kuul* wens wörd go,han morn nomi en test,vellcht chöntsch ned jo morn si met nä (My emphasis)[35]

(311) Hoi [Name]. Han dir grad na schnell es Mail gmacht, wäg dä Adresse fürs Pizzaesse morn! Bis denn, wird sicher *ql*! Grüässli [Name] (My emphasis)[36]

---

31 i caught a cold this week, the climate is highly varied here, that is why i've got quite a bad cough. so I didn't go to work today so as not to infect the babies.

32 copol monday in the palais is one of the absolutely coolest parties. you have to go there. i'll go there again, too, once i've finished my military training school.

33 Now a lazy manikin is going to sleep very tightly! ;-) i've completely dived into the world of work today!! By the way, I don't have to work the next 2 saturdays… Kisses

34 Oh [Name]…i'm just working like mad…i'll get to you soon…an Oringles is hissing, too…grumble

35 i wonder whether you still have a test from [name],the topic is thermodynamics!would be cool if you had one,i've got a test tomorrow afternoon,could you take it with you tomorrow?

36 Hello [Name]. I've just written a mail to you concerning the address of tomorrow's pizza dinner! See you then, it'll certainly be cool! Love [Name]

(312) han no mini prüefige gha, esch voll *isi*, äini hani voll versaut aber esch mer egal ond di ander esch voll *iiiiiiiiiiiiiiisiiiiiii* gsi :) (My emphasis)[37]

On average, Germanised Anglicisms occur 1.7 times per 1000 words, i.e. 8% of all Anglicisms are Germanised ones. Slightly less frequent but still prevalent are the so-called hybrid Anglicisms which combine English and German words in compounds such as *lärn-flow* (313), *Teenie-Liebeskram* (314), and *family-technisch* (315).

(313) Hallo mausebär! bin grad voll im *lärn-flow*! ich blieb no chli i mim 2.diheime. freu mi abr ez scho ufs heicho is bängginäschtli! löf iu so, knutschkuss, [name] (My emphasis)[38]

(314) So, jetzt muß ich mal weiterbügeln, außerdem hab ich Dir ja mal wieder genug *Teenie-Liebeskram* geschrieben u dir die Ohren vollgesülzt.;-) (My emphasis)[39]

(315) Für das darauf folgende We hatten wir ja Samstag abend ins Auge gefasst. Kann dir jetzt auch noch nicht ganz sicher zusagen, da ich vielleicht *family-technisch* etwas eingespannt bin. (My emphasis)[40]

Hybrid Anglicisms in words other than compounds are rare. A typical category of hybrid Anglicisms are particle verbs as *hochsprintet* in example (316). Occasionally, the word *greezli* (also spelt *greetzli*) is used (317) which constitutes a blend of the English wort *greetz* and the Swiss-German word *Grüessli*. On average, hybrid Anglicisms appear 1.26 times per 1000 words, i.e. 6% of all Anglicisms are hybrid ones.

(316) Ein gesundes Haus, in dem er sicherlich nur noch gesunde Nahrung zu sich nehmen wird und die drei Stockwerke des "Frischlufttreppenhauses" immer *hochsprintet*!!!! (My emphasis)[41]

---

37 i've had my tests, it was totally easy, one of them i cocked up completely but I don't mind and the other one was totally easy :)

38 Hey honey! im in the middle of a learning-flow! i'll stay at my second home for a while. but im already looking forward to coming home into your bed! love you so much, snogkiss, [name]

39 Well, i have to go on ironing, anyway, i've again told you enough of my teenie love stories and filled your ears with them.;-)

40 For the ensuing weekend, we have considered Saturday evening. But I still cannot tell you for sure as I may family-technically be occupied.

41 A healthy house in which he will certainly eat healthy food only and always sprint up the three floors of the "fresh-air-staircase"!!!!

(317) Aso bis bald… *Greeezzzllliii* [Name] ;-) see u (My emphasis)[42]

Some Anglicisms also appear to be pseudo-borrowings. These are Anglicisms which are not actually borrowed from the English language, but have been created with English means of word formation within the German language. Examples of such pseudo-Anglicisms are words like *handy* (meaning 'mobile phone', 318), *chips* (meaning 'crisps', 319), or *checken* (meaning 'to understand', 320). Pseudo-Anglicisms, however, are the rarest of all Anglicisms: they only appear 0.59 times per 1000 words on average, which corresponds to 3%.

(318) Hatte mein *Handy* heute vergessen. Mein ab ist beleidigt. Telefonierst du gerade? (My emphasis)[43]

(319) wän: abem 8i bi mir (han chli schinkegipfeli und *chips* und so kauft, aber git nöd wükli znacht) (My emphasis)[44]

(320) Schon alles castellano vergessen oder was? Das ist aber besonders schlecht. Ich *checks* jetzt wieder ohne Ende. Jedenfalls war dieses Jahr auch auf jeden Fall von Vorteil, dass ich auch endlich mal abends mit Basken unterwegs war und gar keine andere Chance hatte als spanisch zu reden. (My emphasis)[45]

Schlobinski et al. (2001: 23) find that Anglicisms are especially frequent in salutation sequences, a result which can also be backed by this data: more than a third, i.e. 34%, of all Anglicisms are used for salutation and farewell formulas respectively, 27% of them constituting salutation formulas, 7% of them farewell formulas. The most common salutation formulas are *hey* (16.4%) and *hi* (9.6%), while the most common farewell formulas are *cu* ('see you', 1.4%) and *mb* ('mail back', 1.2%), as can be inferred from Figure 9-1 and Figure 9-2.

42 Ok see you later… Greetz [Name] ;-) see u

43 I've forgotten my mobile phone today. My answering machine is offended. Are you phoning right now?

44 when: from 8pm onwards at my place (I've bought some ham croissants and crisps but there will not be a proper dinner)

45 Already forgotten all the castellano, huh? This is really bad. I completely understand it again. In any case, it was a great advantage this year that i finally went out with Basques and didn't have another chance than to speak Spanish.

Figure 9-1: Anglicisms in Salutation Formulas

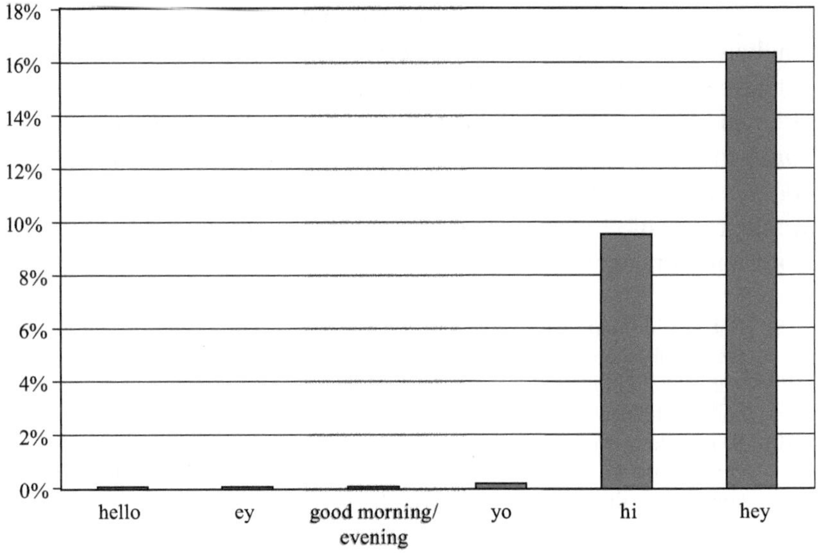

Figure 9-2: Anglicisms in Farewell Formulas

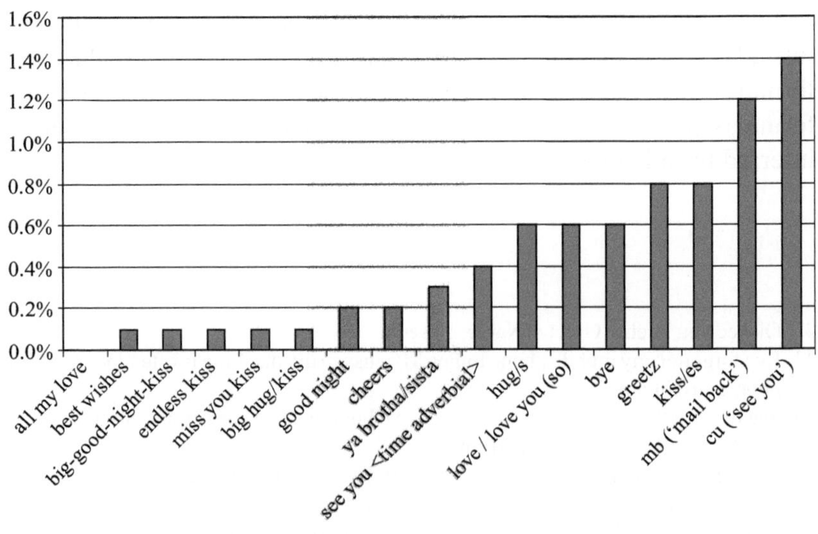

Comparing the English salutation formulas with the German and Swiss-German ones, one finds that there are two kinds of formulas in the latter which do not exist in the English corpora, these being *ey* and *good evening*. The *good-morning*-formula is not a typical salutation formula in English text messages either. Nevertheless, there is one text message in the English SMS corpus which uses this greeting, as example (321) illustrates.

(321) *Good morning*.ah german will be awful,but i bet your spanish paper screamed A* I'm really more than a bit fond of you. xx (My emphasis)

The *ey*- and the *good-evening*-formulas appear to be Helvetic Anglicisms, as they occur only in the Swiss-German corpora. An example of each is provided in quotations (322) and (323).

(322) *EY* ALTER KUMPEL. BESCH HÖT OBE IN LUZERN? (My emphasis)[46]
(323) Hey *good evening* Dear! han der eigentlich nur wellä guet Nacht wünsche und dir sege, das ich mich uf morn freue! Big-good-night-kiss (My emphasis)[47]

Another Helvetic Anglicism is the farewell formula *greetz*. It does not occur in the English (nor German) corpora. Consulting the online *Urban Dictionary*, however, it can be seen that the word has already found entry and is described as another term for 'hello'. Yet, this equivalent constitutes a significant difference to the Swiss-German semantics of the word *greetz*: whereas it is used as a farewell formula in Swiss-German, the word *greetz* is employed as a salutation formula in English. Thus *greetz* seems to be a Helvetic (pseudo-)Anglicism not only on the level of usage, but also in its semantic dimension. As for its etymology, *greetz* derives from the word 'greetings' and has already undergone the *word-final-/s/*-mutation described in Section 3.5.2.3, Non-Conventional Spelling: the word final /s/ in the word *greets* has commonly turned into a word final /z/, and that is why *greetz*, in 81% of the cases, is written with *z*, as depicted in example (324). However, the farewell-formula is

46 HEY OLD CHAP. WILL YOU BE IN LUCERNE TONIGHT?
47 Hey good evening dear! I only wanted to wish you a good night and tell you that im looking forward to tomorrow! Big-good-night-kiss

occasionally also found in other spelling variants such as *greets* (325) or *greez* (326).

(324) aso wünschder na en schöne firtig morn! *greetz* (My emphasis)[48]
(325) Hi [Name]!chume hüt nur is 2. training.falls au nur dänn chunsch,chasch natürli mitfahre.*greets* (My emphasis)[49]
(326) alloä! mer hend di falsche tascherächner! nanana, hesch gspickt! hehe! die 12 sozial-indikatore ;) du, hemmer im rw öbis i dem blaue buech? hdl! *greez*! (My emphasis)[50]

Neither the lexical item *ey* nor *greetz* are found in the *Dictionary of European Anglicisms.* They seem to be new Anglicisms which have originated in the Swiss-German CMC-domain. Most likely they will spread and may soon be found in German messages as well. By the way, the interjection *ey* is affected by *h*-dropping, a rule which has been described in Section 3.5.2.3, Non-Conventional Spelling. This is highly remarkable, for it may be a slight indication that this rule even bears the potential of universal value.

Anglicisms, as this section has shown, are a noteworthy feature of computer-mediated communication. On average, they occur 21.59 times per 1000 words. Compared with emails, they are more common in text messages. The fact that emails only contain half as many Anglicisms as SMS allows one to speculate that they might even be rarer in paper-based non-electronic letter mail. Thus, it appears that the more synchronous the medium is, the more it contains instances of Anglicisms. In other words, the hypothesis is that (near-)synchronous written communication such as text messaging significantly fosters the use of Anglicisms – a tendency which will quite likely become even more pronounced with the further development of mobile communication technology and the growing facility to contact people all over the world easily and cheaply. Hence, with CMC, the world is getting smaller and the English language

48 well I wish you a beautiful holiday tomorrow! greetz

49 Hi [Name]!i'm only coming to the 2nd training today.if you also come to the 2nd training only,i can give you a lift of course.greets

50 hello! we've got the wrong calculators! nanana, you have had a crib! hehe! the 12 social indicators ;)by the way, do we have any homework in accountancy in the blue book? i love you! greez!

is gaining importance so that it intrudes into other languages quite noticeably. At this point it remains mere speculation whether emails will eventually contain more Anglicisms, too, once email technology becomes mobile. Diachronic studies that follow this research will certainly be revealing and bring forth the direction into which the tendency finally leads more clearly.

10 Conclusion

This study presents a number of interesting and new results, which offer a deep insight into the linguistics of computer-mediated communication and lead to a better understanding of it. Yet, the results are not of permanent value because email and SMS as well as MMS technology are (still) developing rapidly. There will certainly be major changes in the near future in these media. Today, email is still stationary, but there is a high demand for wireless email. The advantages of wireless technology are one of the reasons why in Europe SMS is more popular than email (NUA 2002). Certain email services already offer the recipient to be notified about incoming emails via SMS, which is only the beginning of the development of a mobile email technology that will inevitably change communication. SMS, on the other hand, is faced with serious competitors, and it has long been prophesied that it would soon fall into disuse. Despite prophecies of doom, however, the Short Message Service will not be replaced in the near future. In contrast to MMS, which has by far been overestimated, SMS has been underestimated. Nevertheless, it must be recognised that text messaging has serious competitors indeed. An MSN study on Instant Messaging indicates that this service is slowly displacing SMS. The study was conducted between April and May 2004 and interviewed 50 000 users of the MSN Messenger of 13 countries, among them Germany, England, Italy, France, and the USA. Computer Human Interaction expert Helen Petrie from London's City University analysed the results and came to the conclusion that a third of the users in Germany already use the MSN Messenger more often than SMS (*heise online* 2004). This anticipates that once Instant Messaging is available on mobile phones, figures will show a dramatic increase in IM usage. And this will not be long: at the 3GSM World Congress in Barcelona in February 2006, mobile phone providers have decided on an Instant Messaging service which will also work among different networks and can be considered the mobile version of the Internet based IM services. It will enable mobile phone users to directly communicate with each other via the keypads of their mobile phones: upon pressing the enter key, the written text immediately appears on the mobile phone of the intended person. It is assumed that once people can use Instant Messaging via their mobile phones, SMS will slowly be displaced (*20Minuten Online* 2006).

While Instant Messaging is not yet available within the European mobile phone networks, MMS entered the European market four years ago. However, it is still not as popular as one might have expected, which is mainly due to the unappealing pricing scheme as pointed out in Section 8.2, Historical Background. In addition, the critical mass that can send and receive MMS has not been reached, so that those who already have an MMS-enabled mobile phone may have little opportunity to actually use this service. But this will change, as figures and market expectations indicate, and with it, there will be changes in the linguistics as well as in the use of the medium in general. Moreover, today's reservations about these new communication technologies will disappear as it is always the case once a new medium has established itself. In fact, contrary to a hundred years ago when people were worried about not being able to see the person with whom they were speaking while talking on the phone, today's users are rarely concerned with such issues (Baron 1998a: 165). Similarly, computer-mediated communication will undergo pragmatic changes and so, while it is still widely felt that bad news should not be sent via email or SMS, it is possible that such reservations will no longer persist with the growing familiarity and natural use of these electronic media. What is more, the gap of email and SMS/MMS usage between older and younger generations will decrease, for today's young people will be tomorrow's adults, and for them, these media will be an unquestioned means to organise their lives. Consequently, email and SMS will increasingly gain acceptance, also in those parts of society where they have not been widely used yet, and thus will become more common even for those messages which would, at this stage, not be sent via these media.

It is obvious that major changes will take place within CMC and it will certainly be a great benefit to compare succeeding studies with the data of this book in order to see how fast and in which direction the (linguistic) changes ultimately lead us. So far, however, there is no comparable data at hand, as this volume is innovative for various reasons. First, there has never been a comparison of private emails and text messages in the Anglophone area. Second, single text messages and linked text messages have not been analysed separately and put in relation to each other as yet. Third, there is no corpus-based study which discusses the differences of text messages and telegrams. Fourth, multimedia messages have so far only been analysed on the basis of design studies and not on the basis of actual corpora. And finally, Anglicisms have never been more closely scrutinised in private German and Swiss-German emails and text messages. In this respect, this volume provides various new results, all offering a more profound understanding of the (linguistic) nature of emails, text and multimedia messages.

The study has demonstrated that email and text messages are linguistically different modes of communication. Both are characterised by a conceptually oral style, whereby SMS makes greater use of the analysed features than email. In contrast to text messages, the writing of emails does not significantly deviate from what one would expect to find in informal letters, postcards or memoranda, and so it can be concluded that "participants are using fairly standard written English in all its aspects – spelling, punctuation, grammar, general style" (Petrie 1999: 27–29). In other words, "the use of a new 'email style' was not strong, but there is evidence that people are innovating in the way they use the written format to try to communicate some of the emotional content of their message" (Petrie 1999: 27–29). To actually work out the differences between informal letters and emails, a comparison between private emails and personal letters would be required. But this is, in fact, a challenging task for two reasons: first, people rather phone or communicate via electronic media than compose letters, which makes it difficult to acquire a reasonably sized corpus of private letters and second, private letters have been widely neglected in linguistic studies so that little research material is at hand. Nevertheless, a linguistic comparison between letters and emails would be worthwhile and could show more accurately how far these genres actually differ.

This volume has further worked out some of the characteristics of email and SMS. There are, however, various other features which would be worth examining, especially in order to define the difference between emails and SMS more closely. These include an analysis of the lexical density as well as examinations of elements that were not researched in this study, e.g. the use of negations, hedges, amplifiers, inflectives, etc. Considering the continuous development of these registers, a contrastive diachronic study of email and text messages would help us to see how quickly CMC is changing. Moreover, recent studies have always focused on the sender of a message and not on the receiver. In other words, the question has always been how the messages are composed but not how they are read. Following Halliday's statement, however, that "what counts as literacy has to change as our reading and writing practices change in response to ever more complex and dynamic developments in communication technology" (Halliday 1978 quoted from Maybin and Mercer 1996: 82), it would be interesting to analyse how the reading, i.e. the perception of text has changed in order to see whether the hypothesis that there is a correspondence between quick composition and cursory reading proves true. Also, the reader is, for example, significantly more tolerant towards spelling mistakes and erroneous language for two reasons: first, it is often difficult in text-speak to actually categorise instances of non-standard spelling as mistakes and second, erroneous language is not regarded as a matter of bad writing skills, but related to time constraints and lack of space. Outside the electronic medium, spelling mistakes would inevitably lead to the assumption that the author does not know the correct orthography or, depending on how many mistakes there are, might not be educated. In CMC, in contrast, it is taken for granted that the user knows about the correct spelling, but consciously disregards the traditional grammar and spelling rules. Since mistakes do not matter in electronic media, people are again encouraged to compose 'letters', and as email and SMS offer great freedom and do not prescribe rules, 'letter-writing' is accessible to almost anyone. In other words, even those people with poor writing abilities are encouraged to communicate in writing. In this context, Kasesniemi and Rautiainen (2002: 284) observe that "boys would write even less than they do now without SMS and email".

In summary, this study points out some of the characteristics of email and SMS. It compares the two modes of communication with each other and concludes that neither of the two registers can be positioned along the written/spoken continuum, for they also contain elements that are beyond it. Moreover, it shows that many of the characteristic features of email and SMS are not actually new and not necessarily restricted to

CMC. It can thus be assumed that it is not the quality of the linguistic elements used in these modes of communication that have changed, but their quantity. In other words, the linguistic changes are a matter of degree: whereas abbreviations, for instance, have been known for 150 years or more, they have never been as widely used as in these new modes of interaction.

The study has further brought to light the differences between single text messages and linked ones and shown that single SMS are considerably more dialogic in nature than linked text messages. Close analysis of these two types of text messages has also proved what has long been assumed by linguists, namely that the restriction of length is not the only factor responsible for the reduced sentence style of text messages. While the restricted number of characters available in text messages may initially have led to a more extensive use of abbreviations on both the lexical as well as the syntactic level, other factors such as the cumbersome typing as well as the need for brevity and speed and the factor of immediacy have equally contributed to the reduced register, so that by now, the creation of reduced sentences seems to have become a marker of SMS communication regardless of message length. What is more, the lexical abbreviations often constitute simplifications towards a "more 'authentic' representation of speech" (Thurlow 2003). And since text messages are close to conceptual orality, these abbreviations are a logical conclusion and have little to do with the actual length of text messages.

A further comparison with telegrams proved to be worthwhile especially with regard to the different reduction practices: while text messages make use of both lexical and syntactic abbreviations, the language of telegrams appears to be marked by syntactic reductions only. Therefore, it can be summarised that there are certain parallels as well as differences. Yet, it is problematic to speak of a renaissance of telegrams when it comes to text messaging, for telegrams are by far less dialogic and show fewer oral qualities than SMS. Moreover, one of the most significant characteristics of text messaging, i.e. its ludic features, are totally absent in telegrams. Hence the only justification to speak of a renaissance of telegrams is given by the syntactic ellipses – an element which is even more prevalent in telegrams than in text messages. Apart from that, however, there are hardly any parallels, so that it is inappropriate to equate text messages with telegrams.

Unfortunately, not enough English multimedia messages could be gathered for this study so as to be able to base the analysis exclusively on an English corpus. In 2006, mobile multimedia still appeared to be too small a phenomenon to be linguistically studied with statistical means. The analysis of the multimedia messages of the German and Swiss-German corpora, however, has brought about some revealing results concerning the use of this service. It has shown that multimedia messages are considerably less dialogic than text messages and are often used to share one's place or private sphere. Multimedia messages are usually sent among strong ties, i.e. among close friends or family members, and often contain snapshots of people. They can be performative in nature and as such generate an act, informative and simply provide some pieces of information, or else, they may constitute problem-solving messages which help the sender to fulfil some task. Yet, people are still reluctant when it comes to multimedia messaging and prefer other channels for communication, especially text messaging, for SMS is cheaper, usually quicker, and generally more reliable.

The analysis of the pervasiveness of Anglicisms in German and Swiss-German emails and text messages has clearly shown that Anglicisms are especially prevalent in text messages. In fact, it seems that texting has fostered the use of English vocabulary within the German domain, and it is indeed remarkable how easily German and Swiss-German texters code-switch to English vocabulary. There are four different types of Anglicisms, namely pseudo-Anglicisms, hybrid Anglicisms, Germanised Anglicisms, and genuine Anglicisms. The latter type is most commonly used, in simple elements as well as in compounds, phraseologisms, or initialisms. Anglicisms are especially frequent in salutation formulas: almost a third of all Anglicisms constitute greeting sequences.

An interesting question that cannot be answered at this point is how fast and to what extent text-speak will be acknowledged as a linguistic variety of communication. Will there be new spelling conventions that may, at some further stage, also be accepted in other genres? And with the continuing tendency towards orality, will established grammar/spelling rules for unplanned discourse emerge that find entry into grammar books? These questions remain open and the answer partly depends on how computer technology will change in the future. SMS will decline, as has been predicted, and will be overtaken by newer technologies such as multimedia messaging, which enables the user to add pictures and sounds to the text. This implies that the language of text-speak will not only become more visual, but even audiovisual. Multimedia messages may also contain the sender's own voice. Such voice-messages could eventually be compared to messages on answering machines. The advantage of silent communication would be lost in this scenario, but Instant Messaging might by then be the preferred medium to communicate silently.

From a linguistic point of view, there will be an even greater area of research in the future and the most interesting studies are yet to come, namely the diachronic ones which report on the quick changes in electronically composed writing. Language continues to change; it has always been changing but never as quickly as now. In this respect, language is also a mirror of society: the computer has made life faster and thus greater changes are taking place within less time – not only socially, but also linguistically.

References

20Minuten Online. 2006. "Instant Messaging anstelle von SMS." 16 February 2006. <http://www.20min.ch/tools/suchen/story/23849904> [15/02/06].

ADC Telecommunications. 1999. "NewNet Smserver™: Wireless Short Message Service Tutorial." <http://www.mobilein.com/sms_tutorial.pdf> [20/12/01].

Altendorf, Ulrike. 2003. *Estuary English. Levelling at the Interface of RP and South-Eastern British English.* Tübingen: Gunter Narr Verlag.

Androutsopoulos, Jannis and Gurly Schmidt. 2001. "SMS-Kommunikation: Ethnografische Gattungsanalyse am Beispiel einer Kleingruppe." Manuskript angenommen zur Publikation in der *Zeitschrift für Angewandte Linguistik*, 36/2002. <http://www.ids-mannheim.de/prag/sprachvariation/tp/tp7/sms.htm> [18/07/03].

Androutsopoulos, Jannis and Gurly Schmidt. 2004. "löbbe döch. Beziehungskommunikation mit SMS." In: *Gesprächsforschung – Online-Zeitschrift zur verbalen Interaktion*, 5. 50–71. <http://www.gespraechsforschung-ozs.de> [05/02/06].

Androutsopoulos, Jannis. 1999. "Der Name @." In: *Networx*, 6. <http://www.mediensprache.net/de/networx/docs/networx-6.asp> [31/07/03].

Anonymous 2003. <http://fs.mykindaplace.com/life/life_texting.asp> [12/12/03].

Anonymous. 2006. "SMS Geburtstagssprüche." <http://www.geburtstagsspruche.net/sms-geburtstagsspruche.html> [16/01/06].

Avec Mobile. 2004. "Camera phone etiquette." 14 June 2004. <http://www.avecmobile.com/index.php?id=256> [21/07/06].

Ayrton, William Edward. 1901. "Syntonic Wireless Telegraphy. [Ayrton prediction]." In: *Electrical Review*, 29 June 1901. <http://earlyradiohistory.us/1901ayrt.htm> [12/07/06].

Barber, Charles. 1993. *The English Language. A Historical Introduction.* Cambridge: Cambridge University Press.

Baron, Naomi. 1984. "Computer Mediated Communication as a Force in Language Change." In: *Visible Language*, 18 (2). 118–141.
Baron, Naomi. 1998a. "Letters by phone or speech by other means: the linguistics of email." In: *Language and Communication*, 18. 133–170.
Baron, Naomi. 1998b. "Writing in the Age of Email: The Impact of Ideology versus Technology." In: *Visible Language*, 31 (1). 35–53.
Baron, Naomi. 2000. *Alphabet to Email: How Written English Evolved and Where It's Heading.* New York and London: Routledge.
Barton, David. 1994. *Literacy: An Introduction to the Ecology of Written Language.* Oxford: Blackwell.
Battarbee, Katja and Esko Kurvinen. 2005. "Supporting Creativity – Co-experience in Mobile Multimedia Messaging." In: Leslie Haddon, Enid Mante, Bartolomeo Sapio, Kari-Hans Kommonen, Leopoldina Fortunati and Annevi Kant (eds.). *Everyday Innovators: Researching the Role of Users in Shaping ICTs.* Dordrecht: Springer. 71–85.
Bergs, Alexander and Angela Kesseler. 2003. "Literacy and the new media. Vita brevis lingua brevis." In: Jean Aitchison and Diana M. Lewis (eds.). *New Media Language.* London: Routledge. 75–84.
Bergs, Alexander. 2003. "The linguistics of text messaging." Conference Paper from The XVIth International Conference on Historical Linguistics (ICHL 2003) Copenhagen, 11–15 August 2003.
Besnier, Niko. 1993. "Literacy and feelings: the encoding of affect in Nukulaelae letters." In: Brian Street (ed.). *Cross-Cultural Approaches to Literacy.* Cambridge: Cambridge University Press. 62–86.
Beutner, Yvonne. 2002. *E-Mail Kommunikation. Eine Analyse.* Stuttgart: ibidem Verlag.
Biber, Douglas, Stig Johansson, Geoffrey Leech, Susan Conrad and Edward Finegan. 1999. *Longman Grammar of Spoken and Written English.* Essex: Pearson Education Ltd.
Biber, Douglas. 1988. *Variation Across Speech and Writing.* Cambridge: Cambridge University Press.
Biber, Douglas. 1995. *Dimensions of Register Variation: A Cross-Linguistic Perspective.* Cambridge: Cambridge University Press.
Bittner, Johannes. 2003. *Digitalität, Sprache, Kommunikation. Eine Untersuchung zur Medialität von digitalen Kommunikationsformen und Textsorten und deren varietätenlinguistischer Modellierung.* Berlin: Erich Schmidt Verlag.

Brown, Alexandra. 2002. "The language and communication of SMS: An exploratory study of young adults text-messaging." Unpublished BA dissertation, Cardiff University.
Buckingham, Simon. 2001. "Success 4 SMS." <http://www.gsmworld.com/technologies.sms.html> [12/12/01].
Busse, Ulrich. 2004. "Anglizismen in Deutschland: historische Entwicklung, Klassifizierung, Funktion(en) und Einstellungen der Sprachteilhaber." In: Sandro M. Moraldo and Marcello Soffritti (eds.). *Deutsch aktuell. Einführung in die Tendenzen der deutschen Gegenwartssprache.* Rom: Carocci. 81–93.
Cameron, Deborah. 1995. *Verbal hygiene.* London: Routledge.
Campell, Todd. 1998. "The 1st Email Message." In: *Pretext Magazine*, March 1998. <http://www.pretext.com/mar98/features/story2.htm> [26/10/01].
Chafe, Wallace and Jane Danielewitz. 1987. "Properties of Spoken and Written Language." In: Rosalind Horowitz and Jay S. Samuels (eds.). *Comprehending Oral and Written Language.* San Diege [etc.]: Academic Press. 83–113.
Cho, Natasha. Forthcoming. "Linguistic Features of Electronic Mail in the Workplace: A Comparison with Memoranda." In: Susan Herring (ed.). *Computer-Mediated Conversation.*
Chomsky, Noam. 1957. *Syntactic Structures.* The Hague: Mouton.
Christian, Bruce. 2001. "Martin Cooper: Cell Phone 'Father' Squeezes Spectrum." In: *Phono +, The Journal for Wholesalers, Resellers, and Agents of Communications, Computing, and Content Services.* <http://www.phoneplusmag.com/articles/111feat1.html> [03/12/03].
Collot, Milena and Nancy Belmore. 1996. "Electronic Language: A New Variety of English." In: Susan Herring (ed.). *Computer-Mediated Communication: Linguistic, Social and Cross-Cultural Perspectives.* Amsterdam, Philadelphia: John Benjamins Publishing Company. 13–28.
Colombo, Fausto and Barbara Scifo. 2005. "The Social Shaping of New Mobile Devices Among Italian Youth." In: Leslie Haddon, Enid Mante, Bartolomeo Sapio, Kari-Hans Kommonen, Leopoldina Fortunati and Annevi Kant (eds.). *Everyday Innovators: Researching the Role of Users in Shaping ICTs.* Dordrecht: Springer. 86–103.

Crystal, David. 1995. *The Cambridge Encyclopaedia of the English Language*. Cambridge: Cambridge University Press.

Crystal, David. 2001. *Language and the Internet.* Cambridge: Cambridge University Press.

Crystal, David. 2002. *the english language. a guided tour of the language.* 2nd edition. London [etc.]: Penguin.

Crystal, David. 2004. *The Language Revolution.* Cambridge: Polity Press.

Cumming, J. Douglas. 1995. "The Internet and the English Language." In: *English Today*, 11 (1). 3–8.

Daft, Richard and Robert Lengel. 1984. "Information richness: a new approach to managerial information processing and organisation design." In: Barry M. Staw and Larry L. Cummings (eds.). *Research in Organisational Behaviour*, 6. 191–233.

Daft, Richard and Robert Lengel. 1986. "Organisational information requirements, media richness and structural design." In: *Management Science*, 32 (5). 554–571.

Dimmick, John, Susan Kline and Laura Stafford. 2000. "The Gratification Niches of Personal Email and the Telephone: Competition, Displacement and Complementarity." In: *Communication Research*, 27. 227–248.

Döring, Nicola. 1999. *Sozialpsychologie des Internet. Die Bedeutung des Internet für Kommunikationsprozesse, Identitäten, soziale Beziehungen und Gruppen.* Göttingen: Hogrefe.

Döring, Nicola, Christine Dietmar, Alexandra Hein and Katharina Hellwig. 2005. "Contents, Forms, and Functions of Interpersonal Pictorial Messages in Online and Mobile Communication." Plenary talk held at the Conference on Seeing, Understanding, Learning in the Mobile Age, Budapest, 28–30 April 2005.

Döring, Nicola. 2002. "Kurzm. wird gesendet – Abkürzungen und Akronyme in der SMS-Kommunikation." <http://www.nicola-doering.de/publications/sms-kurzformen-doe-ring-2002.pdf> [12/07/03].

Dürscheid, Christa and Arne Ziegler (eds.). 2002. *Kommunikationsform E-Mail.* Tübingen: Stauffenberg.

Dürscheid, Christa. 2000. "Verschriftungstendenzen jenseits der Rechtschreibreform." In: *Zeitschrift für germanistische Linguistik*, 28. 237–247.

Dürscheid, Christa. 2002a. "SMS-Schreiben als Gegenstand der Sprachreflexion." In: *Networx*, 28. <http://www.mediensprache.net/networx/networx-28.pdf> [08/08/03].

Dürscheid, Christa. 2002b. *Einführung in die Schriftlinguistik*. Wiesbaden: Westdeutscher Verlag (= Studienbücher zur Linguistik 8).

Dürscheid, Christa. 2003a. "Medienkommunikation im Kontinuum von Mündlichkeit und Schriftlichkeit. Theoretische und empirische Probleme." In: *Zeitschrift für Angewandte Linguistik*, 38. 37–56.

Dürscheid, Christa. 2003b. Syntaktische Tendenzen im heutigen Deutsch. Manuskript zur Antrittsvorlesung an der Universität Zürich am 21.6.2003.

Dürscheid, Christa. 2004. "Netzsprache – ein neuer Mythos." In: Michael Beisswenger, Ludger Hoffmann and Angelika Storrer (eds.). *Osnabrücker Beiträge zur Sprachtheorie, 68. Internetbasierte Kommunikation*. Duisburg: Gilles und Francke. 141–157.

Dürscheid, Christa. 2005. "Medien, Kommunikationsformen, kommunikative Gattungen." In: *Linguistik online*, 22 (1). <http://www.linguistik-online.de/22_05/duerscheid.html> [30/01/06].

Eichhoff-Cyrus, Karin M. and Rudolf Hoberg (eds.). 2000. *Die deutsche Sprache zur Jahrtausendwende. Sprachkultur oder Sprachverfall?* Mannheim: Dudenverlag.

Einstein, Albert. "Albert Einstein Quotes." <http://www.humboldt1.com/~gralsto/einstein/quotes.html> [22/06/06].

Fahlman, Scott E. 2002. "Smiley Lore." <http://www-2cs.cmu.edu/~sef/sefSmiley.htm> [07/07/03].

Ferrara, Kathleen, Hans Brunner and Greg Whittemore. 1991. "Interactive Written Discourse as an Emergent Register." In: *Written Communication*, 8 (1). 8–33.

Fielding, Rachel. 2002. "Fastest text in the west." In: *IT WEEK*, 22 November 2002. <http://www.itweek.co.uk/News/1127069> [23/07/03].

Finegan, Edward. 1999. *Language its Structure and Use*. 3rd edition. Fort Worth [etc.]: Harcourt Brace College Publishers.

Freyermuth, Gundolf. 2002. *Kommunikette 2.0*. Heidelberg: Heise.

Frohlich, David and Ella Tallyn. 1999. "Audiophotography: practice and prospects." In: *Proceedings of CHI '99*. New York: ACM SIG-CHI. 296–297.

Georgakopoulou, Alexandra. Forthcoming. "'On for Drinkies?' E-mail Cues of Participant Alignments." In: Susan Herring (ed.). *Computer-Mediated Conversation.*

Gerber, Erika, Konrad Ehlich and Jan-Dirk Müller (eds.). 2002. *Materialität und Medialität von Schrift.* Bielefeld: Aisthesis Verlag.

Gerhardinger, Monika. 2000. *Short Messages: Eine sprachliche Untersuchung im Vergleich zu E-mail und Chat.* Diplomarbeit, Universität Passau.

Görlach, Manfred (ed.). 2001. *A Dictionary of European Anglicisms.* Oxford: Oxford University Press.

Graham, Gordon. 1999. *The Internet: A Philosophical Inquiry.* London: Routledge.

Griffiths, Richard. 2002. "Chapter Three: History of Electronic Mail." <http://www.let.leiden-univ.nl/history/ivh/chap3.htm> [23/11/03].

Grinter, Rebecca E. and Margery A. Eldridge. 2001. "y do tngrs luv 2 txt msg?" In: Wolfgang Prinz, Matthias Jarke, Yvonne Rogers, Kjeld Schmidt and Volker Wulf (eds.). *Proceedings of the Seventh European Conference on Computer-Supported Cooperative Work, ECSCW '01.* Netherlands: Kluwer Academic Publishers. 219–238.

GSM MOU Permanent Secretariat. 1990. *A Guide to Pan-European Digital Cellular Radio MOU-MP.* Document 4, Version 3.1.0, October.

Günther, Ulla and Eva Lia Wyss. 1996. "E-mail-Briefe – eine neue Textsorte zwischen Mündlichkeit und Schriftlichkeit." In: Ernest Hess-Lüttich, Werner Holly and Ulrich Püschel (eds.). *Textstrukturen im Medienwandel.* Frankfurt am Main: Peter Lang GmbH. 61–86.

Haase, Martin, Michael Huber, Alexander Krumeich and Georg Rehm. 1997. "Internetkommunikation und Sprachwandel." In: Rüdiger Weingarten (ed.). *Sprachwandel durch Computer.* Opladen: Westdeutscher Verlag. 51–85.

Haddon, Leslie. 2000. "The Social Consequences of Mobile Telephony: Framing." <http://www.mot.chalmers.se/dept/tso/haddon/Framing.pdf> [12/12/03].

Halliday, Michael. 1978. *Language as Social Semiotic.* London: Arnold.

Halliday, Michael. 1987. "Spoken and written modes of meaning." In: Rosalind Horowitz and S. Jay Samuels (eds.). *Comprehending Oral and Written Language.* Orlando, Florida: Academic Press. 55–82.

Hamill, Lynne and Amparo Lasen (eds.). 2005. *Mobile World. Past, Present and Future.* London [etc.]: Springer.

Haugen, Einar. 1996. "Dialect, Language, Nation." In: *American Anthropologist*, 68. 922–935.

heise online. 2004. "MSN-Studie: Instant Messaging verdrängt SMS." <http://www.heise.de/newsticker/meldung/print/49000> [06/07/06].

Herring, Susan (ed.). Forthcoming. *Computer-Mediated Conversation.*

Herring, Susan. 1996. "Two variants of an electronic message schema." In: Susan Herring (ed.). *Computer-Mediated Communication: Linguistic, Social and Cross-Cultural Perspectives.* Amsterdam, Philadelphia: John Benjamins Publishing Company. 81–106.

Herring, Susan. 2001. "Computer-mediated discourse." In: Deborah Schiffrin, Deborah Tannen and Heidi E. Hamilton (eds.). *The handbook of discourse analysis.* Oxford: Blackwell. 612–634.

Hoberg, Rudolf. 2000. "Sprechen wir bald alle Denglisch oder Germeng?" In: Karin M. Eichhoff-Cyrus and Rudolf Hoberg (eds.). *Die deutsche Sprache zur Jahrtausendwende. Sprachkultur oder Sprachverfall?* Mannheim: Dudenverlag. 303–316.

Höflich, Joachim and Julian Gebhard (eds.). 2003. *Vermittlungskulturen im Wandel. Brief, Email, SMS.* Frankfurt am Main: Peter Lang GmbH.

Höflich, Joachim and Patrick Rössler. 2000. Forschungsprojekt "Jugendliche und SMS. Gebrauchsweisen und Motive." Zusammenfassung der ersten Ergebnisse. <http://www.uni-erfurt.de/km/forschung/smsreport.doc> [18/07/03].

Höflich, Joachim and Patrick Rössler. 2001. "Mobile schriftliche Kommunikation – oder: E-Mail für das Handy. Die Bedeutung elektronischer Kurznachrichten (Short Message Service) am Beispiel jugendlicher Handynutzer." In: *Medien & Kommunikationswissenschaft*, 49 (1). 437–461.

Höflich, Joachim. 2003. "Vermittlungskulturen im Wandel: Brief – Email – SMS." In: Joachim Höflich and Julian Gebhard (eds.). *Vermittlungskulturen im Wandel. Brief, Email, SMS.* Frankfurt am Main: Peter Lang GmbH. 39–61.

Horowitz, Rosalind and Marian G. Barchilon. 1994. "Stylistic Guidelines for Email." In: *IEEE Transactions on Professional Communication.*

Special Issue on Electronic Communication and Interaction, 37 (4). 207–212.

Huff, Charles and Jonathon Rosenberg. 1989. "The on-line voyeur: Promises and pitfalls of observing electronic interaction." In: *Behaviour Research Methods, Instruments, and Computers*, 21. 166–172.

I Will Follow Services™. 2003. "E-Mail Etiquette." <http://www.iwill-follow.com/emailetiquette.html> [13/07/03].

IEEE. 2003. "E-Mail Netiquette Guidelines." <http://elecomm.ieee.org/email-netiquette.shtml> [13/07/03].

Ito, Mizuko and Daisuke Okabe. 2003. "Camera phones changing the definition of picture-worthy." In: *Japan Media Review*. <http://www.ojr.org/japan/wireless/1062208524.php> [14/02/06].

Janney, Richard W. 1996. "E-mail and Intimacy." In: Barbara Gorayska and Jonathon L. Mey (eds.). *Cognitive Technology: In Search of a Human Interface*. Amsterdam: Elsevier Science. 201–211.

Jansen, Erin. 2002. *NetLingo. The Internet Dictionary*. Ojai (CA): Netlingo Inc.

Johnson, Steven. 1997. *Interface Culture. How new technology transforms the way we create and communicate*. San Francisco: Harper.

Johnston, Jenifer. 2003. "Teachers call for urgent action as pupils write essays in text-speak. Children's obsession with mobile-phone messaging is seriously damaging their written English, warn experts." In: *Sunday Herald*, 02 March 2003. <http://www.sundayherald.com/print31826> [08/08/03].

Jucker, Andreas H. and Yael Ziv. (eds.). 1998. *Discourse Markers: Descriptions and Theory*. Amsterdam: John Benjamins Publishing Company.

Kasesniemi, Eija Liisa and Pirjo Rautiainen. 2002. "Mobile culture of children and teenagers in Finland." In: James E. Katz and Mark A. Aakhus (eds.). *Perpetual contact: Mobile communication, private talk, public performance*. Cambridge: Cambridge University Press. 170–192.

Kasesniemi, Eija Liisa. 2003. *Mobile Messages: Young People and a New Communication Culture*. Tampere: Tampere University Press.

Katz, James E. and Mark A. Aakhus (eds.). 2002. *Perpetual contact: Mobile communication, private talk, public performance*. Cambridge: Cambridge University Press.

Kiesler, Sara and Lee Sproull. 1986. “Reducing social context cues: Electronic mail in organisational communication.” In: *Management Sci.* 32 (11). 1492–1512.

Kiesler, Sara and Lee Sproull. 1991. *New Ways of Working in the Networked Organization.* Cambridge, Massachusetts: Massachusetts Institute of Technology.

Kiesler, Sara, Jane Siegel and Timothy W. McGuire. 1984. “Social psychological aspects of computer-mediated communication.” In: *American Psychologist*, 39. 1123–1134.

Kiesler, Sara. 1987. “Social aspects of computer environments.” In: *Social Science*, 72. 23–28.

Kindberg, Tim, Mirjana Spasojevic and Fleck Rowanne. 2004. “How and Why People Use Camera Phones.” <http://www.hpl.hp.com/techreports/2004/HPL-2004-216.pdf> [14/02/06].

Kindberg, Tim, Mirjana Spasojevic, Fleck Rowanne and Sellen Abigail. 2005. “The Ubiquitous Camera: An In-Depth Study of Camera Phone Use.” In: *IEEE Pervasive Computing. Mobile and Ubiquitous Systems*, 4 (2). 42–50.

Kinsley, Michael. 1996. “The Morality & Metaphysics of Email.” In: *Forbes ASAP: The Big Issue.* 02 December 1996. 113. <http://www.hwgs.de/e-mail/e_mail/e_mail_06.html> [20/12/01].

Koch, Peter and Wulf Oesterreicher. 1985. “Sprache der Nähe – Sprache der Distanz. Mündlichkeit und Schriftlichkeit im Spannungsfeld von Sprachtheorie und Sprachgeschichte.” In: *Romanistisches Jahrbuch* 36. 15–43.

Koch, Peter and Wulf Oesterreicher. 1994. “Schriftlichkeit und Sprache.” In: Hartmut Günther and Otto Ludwig (eds.). *Schrift und Schriftlichkeit. Handbücher für Sprach- und Kommunikationswissenschaft*, 1. Berlin, New York: Walter de Gruyter. 588–604.

Koskinen, Ilpo and Esko Kurvinen. 2005. “Mobile multimedia and users: On the domestication of mobile multimedia.” In: *Telektronikk*, 3 (4). 60–68. <http://www2.uiah.fi/~ikoskine/recentpapers/mobile_multimedia/telektronikk_koskinen_kurvinen.pdf> [16/02/06].

Koskinen, Ilpo, Esko Kurvinen and Turo-Kimmo Lehtonen. 2002. *Professional Mobile Image.* Finnland: IT Press.

Koskinen, Ilpo. 2004. “Seeing with Mobile Images: Towards Perpetual Visual Contact.” Proceedings of the conference ‘The Global and the Local in Communication: Places, Images, People, Connections', Budapest, 10–12 June 2004. <http://www.fil.hu/mobil/2004/Koskinen_webversion.pdf> [14/02/06].

Koskinen, Ilpo. 2005a. "Mobile Multimedia in Society: Uses and Social Consequences." <http://www2.uiah.fi/~ikoskine/recentpapers/mobile_multimedia/mobilemultimediainsociety.pdf> [15/02/06].

Koskinen, Ilpo. 2005b. "Pervasive Image Capture and Sharing: Methodological Remarks." Workshop at Ubicomp 11–14 September 2005, Tokyo, Japan. The Seventh International Conference on Ubiquitous Computing. <http://www.spasojevic.org/pics/PICS/Ubicomp-05WS2Koskinen.pdf> [16/02/05].

Kuhn, Michael. 2005. "Umsätze mit MMS ziehen an." In: *Handelszeitung*, 29 June 2005. 3.

Kümmel, Bianca. 1997. "Welche Vorteile bietet Email?" In: *hwgs. Kommunikationsmedien*. May 1998. <http://www.hwgs.de/e-mail/e_mail/e_mail_06.html> [15/06/01].

Ladefoged, Peter. 1993. *A Course in Phonetics*. Fort Worth [etc.]: Harcourt Brace College Publishers.

Lan, Li. 2000. "Email: a challenge to Standard English?" In: *English Today*, 16 (4). 23–55.

Lee, Judith Yaross. 2003. "Charting the Codes of Cyberspace: A Rhetoric of Electronic Mail." In: Lance Strate, Ron L. Jacobson and Stephanie B. Gibson (eds.). *Communication and Cyberspace*. 2nd edition. Cresskill: Hampton Press Inc. 307–328.

Lenhart, Amanda, Mary Madden and Paul Hitlin. 2005. "Teens and Technology. Youth are leading the transition to a fully wired and mobile nation." Washington: Pew Internet and American Life Project. <http://www.pewinternet.org/pdfs/PIP_Teens_Tech_July-2005web.pdf> [25/06/06].

Leung, Louis and Ran Wei. 2000. "More than just Talk on the Move: Uses and Gratifications of the Cellular Phone." In: *Journalism and Mass Communication Quarterly*, 77 (2). 308–320.

Licklider, Joseph C. R. and Albert Vezza. 1978. "Applications of Information Networks." In: *Proceedings of the IEEE*, 66 (11). 44.

Light, Ann. 2005. "Mobile Age: The Stories that Pictures tell." <http://www.usabilitynews.com/news/article2568.asp> [14/02/06].

Linfield College. 2003. "E-Mail 'Netiquette'." <http://www.linfield.edu/policy/netiquette.html> [13/07/03].

Ling, Richard and Birgitte Yttri. 1999. "Nobody sits at home and waits for the telephone to ring: Micro and hyper-coordination through the use of the mobile phone." Technical Report 30/1999, Telenor. <http://www.telenor.no/fou/program/nomadiske/articles/08.pdf> [21/07/06].

Ling, Richard and Birgitte Yttri. 2002. "Hyper-coordination via mobile phones in Norway." In: James E. Katz and Mark A. Aakhus (eds.). 2002. *Perpetual contact: Mobile communication, private talk, public performance*. Cambridge: Cambridge University Press. 139–169.

Lingo2Word. <http://www.lingo2word.com/index.php> [10/07/03].

Longman Dictionary of Contemporary English. 1995. Essex: Longman Group Ltd.

Longmate, Elizabeth and Chris Barber. 2002. "A Comparison of Text Messaging and Email Support for Digital Communities: A Case Study." In: Xristine Faulkner, Janet Finlay and Françoise Détienne (eds.). *People and Computers XVI – Memorable Yet Invisible. Proceedings of HCI 2002*. London [etc.]: Springer. 69–88. <http://postgrad.eee.bham.ac.uk/longmate/hci2002.doc> [03/04/06].

Mäkela, Anu, Verena Giller, Manfred Tscheligi and Reinhard Sefelin. 2000. "Joking, storytelling, artsharing, expressing affection: A field trial of how children and their social network communicate with digital images in leisure time." In: Thea Turner, Gerd Szwillus, Mary Czerwinski and Fabio Paterno (eds.). *CHI 2000 Conference proceedings: CHI 2000 – the future is here. Conference on Human Factors in Computing Systems*. New York: ACM Press. 548–555.

Mander, Gabrielle (ed.). 2000. *ltle bk of txt msgs*. London: Michael O'Mara Books Ltd.

Markus, M. Lynne. 1987. "Chargeback as an implementation tactic for office communication systems." In: *Interfaces*, 17. 54–63.

Markus, M. Lynne. 1990. "Towards a 'critical mass' theory of interactive media." In: Janet Fulk and Charles Steinfield (eds.). *Organizations and Communication Technology*. London: Sage. 194–218.

Maybin, Janet and Neil Mercer. 1996. *Using English: From conversation to canon*. London: Routledge.

Mayhew, Anthony Lawson and Walter William Skeat. 1978. A *Concise Dictionary of Middle English from A.D. 1150 to 1580*. Oxford: Clarendon Press.

Maynor, Natalie. 1994. "The Language of Electronic Mail: Written Speech?" In: Greta D. Little and Michael Montgomery (eds.). *Centennial Usage Studies*. Tuscaloosa: University of Alabama. 48–54.

McCormick, Naomi B. and John W. McCormick. 1992. "Computer Friends and Foes: Content of Undergratuates' Electronic Mail." In: *Computers in Human Behaviour*, 8. 379–405.

Meise-Kuhn, Katrin. 1998. "Zwischen Mündlichkeit und Schriftlichkeit: Sprachliche und konversationelle Verfahren in der Computerkommunikation." In: Alexander Brock and Martin Hartung. *Neuere Entwicklung in der Gesprächsforschung. Vorträge der 3. Arbeitstagung des Pragmatischen Kolloquiums Freiburg*. Tübingen: Gunter Narr Verlag. 213–235.

Menon, Naveen, Mark Page, Matthew Watt and Simon Bell. 2005. "Mobinet 2005." An A.T. Kearney/University of Cambridge study. A Selection of Key Findings. October 2005. <http://www.atkearney.com/shared_res/pdf/Mobinet_2005_Detailed_Results.pdf> [23/06/06].

Meyers, Alfred S. 1994. *Letters for all occasions*. New York: Harper.

Milroy, James and Lesley Milroy. 1985. *Authority in Language. Investigating Language Prescription and Standardisation*. London, New York: Routledge.

Mobile Data Association (MDA). 2003. <http://www.mda-mobiledata.org/resource/hottopics/sms.asp> [14/01/04].

Mooney, Elizabeth. 2001. *Daily News – The Independent Source for Wireless Industry News*. 15 January 2001. <http://www.rcrnews.com/rcr/fetch.php3?id=13463> [20/12/03].

Moraldo, Sandro and Marcello Soffritti (eds.). 2004. *Deutsch aktuell. Einführung in die Tendenzen der deutschen Gegenwartssprache*. Rom: Carocci.

Moraldo, Sandro. 2002. "Short Message Service (SMS) oder Zur Linguistik der Kurznachrichten." In: *Sprachspiegel*, 2. 155–166.

Moraldo, Sandro. 2004a. "Linguistische Forme(l)n in textbasierten SMS-Dialogen. Zur kontrastiven Analyse deutscher und italienischer Kurznachrichten." In: Werner Roggausch (ed.). *Germanistentreffen*

Deutschland – Italien, 8.–12.10.2003. Dokumentation der Tagungsbeiträge. Bonn (DAAD). 285–301.

Moraldo, Sandro. 2004b. "Medialität und Sprache. Zur Verlagerung von Sprachkommunikation auf Datentransfer am Beispiel von SMS und eMail." In: Sandro M. Moraldo and Marcello Soffritti (eds.). *Deutsch aktuell. Einführung in die Tendenzen der deutschen Gegenwartssprache*. Rom: Carocci. 253–270.

Moran, Charles and Gail Hawisher. 1998. "The rhetorics and languages of electronic mail." In: Ilana Snyder (ed.). *Page to Screen*. London: Routledge. 80–101.

MORI and Lycos UK. 2000. "I just text to say I love you." Press release, 5 September 2000. <http://www.mori.com/polls/2000/lycos.shtml> [18/07/03].

Murray, Denise. 1990. "CMC." In: *English Today*, 6 (3). 42–46.

Negroponte, Nicholas. 1995. *being digital*. London: Hodder and Stoughton.

Nolde, Dirk. 2001. "Des Präsidenten letzte E-Mail." In: *Die Welt*. 21 March 2001.

NUA. 2002. <http://www.nua.ie> [17/06/03].

Ochs, Elinor. 1979. "Planned and Unplanned Discourse." In: Talmy Givon (ed.). *Syntax and Semantics 12: Discourse and Syntax*. New York: Academic Press. 51–80.

Okabe, Daisuke. 2004. "Emergent Social Practices, Situations and Relations through Everyday Camera Phone Use." Paper presented at Mobile Communication and Social Change, the 2004 International Conference on Mobile Communication in Seoul, Korea. <http://www.itofisher.com/mito/archives/okabe_seoul.pdf> [14/02/06].

Okabe, Daisuke. 2005. "Social practice of Camera Phone in Japan." <http://www.spasojevic.org/pics/PICS/okabe_ubicomp.pdf> [14/02/06].

ONE News. 2003. "Teen habits drive text message boom." 22 January 2003. <http://onenews.nzoom.com/onenews_detail/0,1227,163249-1-454,00.html> [20/07/03].

Ong, Walter. 1982. *Orality and Literacy*. London and New York: Routledge.

Orlowski, Andrew. 2004. "Picture messaging – it's worse than you thought." <http://www.theregister.co.uk/2004/09/15/mms_flop_needs_fix> [15/02/06].

Ortner, Lorelies. 2002. “SMS-Botschaften: Texttypologie aus der Perspektive der SMS-Ratgeberliteratur.” In: Ulrich Schmitz and Eva Lia Wyss (eds.). *Osnabrücker Beiträge zur Sprachtheorie, 64. Briefkommunikation im 20. Jahrhundert.* Duisburg. 207–238.

PC Magazine. 2002. “The Mobile Message.” 21 May 2002. <http://www.ebscohost.com> [13/07/03].

Petrie, Helen. 1999. “Writing in cyberspace: a study of the uses, style and content of email.” Unpublished Paper sponsored by MSN Microsoft.

Pew Internet and American Life Project. 2003. <http://www.pewinternet.org/reports/chart.asp?img=Daily_Activities_11.7.03.htm> [03/12/03].

Pollack, Andrew. 1996. “Happy in the east (--) or smiling :-) in the west.” In: *The New York Times, Late Edition – Final*, 12 August 1996. <http://www.cs.nyu.edu/ms_students/cera7013/class/japan.htm> [12/12/03].

Pritchard, Kate. 2001. “Text messaging comes of age.” In: *IT WEEK*, 25 October 2001. <http://www.itweek.co.uk/Features/1126659> [08/07/03].

Resources Network. 2003. <http://www.cplusplis.com/doc/papers/ascii.html> [28/11/03].

Rice, Ronald E. and Donald Case. 1983. “Electronic Message Systems in the University: A Description of Use and Utility.” In: *Journal of Communication*, 33. 131–152.

Rudy Ian A. 1996. “A critical review of research on electronic mail.” In: *European Journal of Information Systems*, 4 (4). 198–213.

Saberi, Kouroshs and David R. Perrot. 1999. “Cognitive Restoration of Reversed Speech.” In: *Nature*, 29 April 1999. 760.

Sacks, Harvey. 1994. *Lectures on Conversation.* Oxford: Blackwell. 2 Vols.

Schlobinski, Peter, Jens Runkehl and Torsten Siever. 1988. *Sprache und Kommunikation im Internet.* Opladen, Wiesbaden: Westdeutscher Verlag.

Schlobinski, Peter, Nadine Fortmann, Olivia Gross, Florian Hogg, Frauke Horstmann and Rena Theel. 2001. “Simsen. Eine Pilotstudie zu sprachlichen und kommunikativen Aspekten in der

SMS-Kommunikation." In: *Networx*, 22. <http://www.websprache.net/networx/docs/networx-22.htm> [14/09/02].

Schlobinski, Peter. 2000. "Chatten im Cyberspace." In: Karin Eichhoff-Cyrus and Rudolf Hoberg. (eds.). 2000. *Die deutsche Sprache zur Jahrtausendwende. Sprachkultur oder Sprachverfall?* Mannheim: Dudenverlag.

Schlobinski, Peter. 2001a. "*knuddel – zurueckknuddel – dich ganzdollknuddel*." In: Helmut Henne, Angelika Linke, Sigurd Wichter and Herbert Ernst Wiegand (eds.). *Zeitschrift für germanistische Linguistik*. Berlin, New York: Walter de Gruyter. 192–218.

Schlobinski, Peter. 2001b. "Rezension on David Crystal's Language and the Internet." <http://www.websprache.net/literatur/rezensionen/0521802121.pdf> [02/02/06].

Schlobinski, Peter. 2002. "Simsen." Interview published in the online newspaper *dieUniversität.at*. <http://www.fbls.uni-hannover.de/sdls/schlobi/splitter/simsen.htm> [16/07/03].

Schotzger, Erwin. 2002. "SMS vor verspätetem Boom in USA." In: *Telephia*, 17 December 2002. <http://www.innovations-report.de/html/berichte/kommunikation_medien/bericht15292.html> [20/01/06].

Schwitalla, Johannes. 2002. "Kleine Botschaften – Telegramm – und SMS-Texte." In: Ulrich Schmitz and Eva Lia Wyss (eds.). *Osnabrücker Beiträge zur Sprachtheorie, 64. Briefkommunikation im 20. Jahrhundert.* Duisburg. 33–56.

Science and Technology Dictionary. McGraw-Hill Dictionary of Scientific and Technical Terms. 6th edition. <http://www.answers.com/library/Science%20and%20Technology%20Dictionary-cid-1247853054> [18/04/07].

Scifo, Barbara. 2005. "The Domestication of Camera-Phone and MMS Communication. The Early Experiences of Young Italians." In Kristof Nyíri. (ed.). *A Sense of Place. The Global and the Local in Mobile Communication.* Vienna: Passagen Verlag. 363–374. <http://21st.century.phil-inst.hu/Passagen_engl4_Scifo.pdf> [16/02/06].

Seasonal Web. 1999. "Overview of Email History." <http://seasonal.net/WebTools/index.cfm> [18/12/2001].

Segerstad, Ylva Hård. 2002. *Use and Adaptation of Written Language to the Conditions of Computer-Mediated Communication.* (Diss. abstract). Department of Linguistics, Göteborg University, Sweden.

Severinson Eklundh, Kerstin. 1986. *Dialogue processes in computer-mediated communication: the study of letters in the COM system.* (Diss.) Linköping Studies in Arts and Science 6. University of Linköping, Sweden.

Severinson Eklundh, Kerstin and Clare Macdonald. 1994. "The Use of Quoting to Preserve Context in Electronic Mail Dialogues." In: *IEEE Transactions on Professional Communication, Special Issue on Electronic Communication and Interaction*, 37 (4). 197–202.

Severinson Eklundh, Kerstin. Forthcoming. "To Quote or Not to Quote: Setting the Context for Computer-Mediated Dialogues." In: Susan Herring (ed.). *Computer-Mediated Conversation.*

Shapiro, Norman Z. and Robert H. Anderson. 1985. *Toward an Ethics and Etiquette for Electronic Mail.* Santa Monica (CA): Rand Corporation. <http://www.rand.org/publications/MR/R3283> [10/11/03].

Sherwood, Kaitlin Duck. 2001. "A Beginner's Guide to Effective Email." <http://www.webfoot.com/advice/email.top.html?Author> [21/12/01].

Siever, Torsten. 2005. "Von MfG bis cu l8er. Sprachliche und kommunikative Aspekte von Chat, E-Mail." In: *Der Sprachdienst*, 5–6. 137–147.

Snyder, Ilana (ed.). 1998. *Page to Screen: Taking Literacy Into the Electronic Era.* London, New York: Routledge.

Solberg, Are. 2004. *Genres of Communication in Nomadic Work Environments. The Potential of Multimedia Messaging.* Cand. Scient Thesis. <http://folk.uio.no/aresolbe/doku-menter/hovedfag/thesis.doc> [15/02/06].

Solymar, Laszlo. 1999. *Getting the Message. A History of Communications.* Oxford: Oxford University Press.

Spears, Russel, Lea Martin and Tom Postmes. 2001. "Social psychological theories of computer-mediated communication: Social pain or social gain." In: W. Peter Robinson and Howard Giles (eds.). *The New Handbook of Language and Social Psychology.* Chichester: John Wiley and Sons. 601–623.

Spiegel. 2003. "SMS-Sprachalarm an Schulen: 2B or not 2B." 3 March 2003. <http://www.spiegel.de/uni-spiegel/studium/0,1518,238539,00.htm> [10/10/03].

Spitzer, Michael. 1986. "Writing style in computer conferences." In: *IEEE Transactions of Professional Communication*, 29 (1). 19–22.

Srivastava, Laura. 2004. "Social and Human Considerations for a More Mobile World." Background Paper. ITU/MIC Workshop on Shaping the Future Mobile Information Society, Seoul, 4–5 March 2004. <http://www.itu.int/osg/spu/ni/futuremobile/SocialconsiderationsBP.pdf> [15/05/07].

Stein, Dieter. 2001. "Textlinguistik." In: Ansgar Nünning (ed.). *Metzler Lexikon. Literatur- und Kulturtheorie*. 2nd revised and extended edition. Stuttgart, Weimar: Verlag J. B. Metzler.

Stickel, Gerhard. 2004. "Das heutige Deutsch: Tendenzen und Wertungen." In: Sandro M. Moraldo and Marcello Soffritti (eds.). *Deutsch aktuell. Einführung in die Tendenzen der deutschen Gegenwartssprache*. Rom: Carocci. 11–32.

Stroh, Kassian. 2004. "Die Sprache der Jugend." <http://www.sueddeutsche.de/panorama/artikel/883/31852> [11/08/05].

Tannen, Deborah. 1982. *Spoken and Written Language: Orality and Literacy*. Norwood, New Jersey: Albex Publishing Corporation.

Taylor, Alex S. and Richard Harper. 2002. "Age-old practices in the 'New World': A study of gift-giving between teenage mobile phone users." In: *Proceedings of the SIGCHI Conference on Human Factors in Computing Systems*. 439–446. <http://research.microsoft.com/~ast/files/CHI_2002.pdf> [15/02/06].

Taylor, Alex S. and Jane Vincent. 2005. "An SMS History." In: Lynne Hamill and Amparo Lasen (eds.). 2005. *Mobile World. Past, Present and Future*. London [etc.]: Springer. 75–91.

TelegramsOnline. <http://www.telegramsonline.co.uk/aboutus.asp> [22/06/06].

Texting Dictionary. 2004. HarperCollins. <http://www.collins.co.uk/wordexchange/Sections/TextinWords.aspx?letter=A&pg=11> [19/06/06].

The Economist. 2002. "More Power to the Thumb." 22 June 2002. <http://www.economist.com/science/tq/displayStory.cfm?story_id=1176079> [25/08/03].

The Economist. 2003. "No text please, we're American." 5 April 2003. <http://www.economist.com/business/displayStory.cfm?story_id=1683713> [25/08/03].

Thurlow, Crispin. 2001. "Language and the Internet." In: Rajend Mesthrie and Ronald E. Asher (eds.). *The concise encyclopedia of sociolinguistics.* Oxford [etc.]: Elsevier Science Ltd. 287–289.

Thurlow, Crispin. 2002. "Generation Txt." <http://www.shu.ac.uk/daol/articles/open/2002/003/thurlown2002003-02.html> [10/07/03].

Thurlow, Crispin. 2003. "Generation Txt? The sociolinguistics of young people's text-messaging." In: *Discourse Analysis Online.* <http://www.shu.ac.uk/daol/articles/v1/n1/a3/thurlow2002003-paper.html> [30/01/06].

Trask, Robert Lawrence. 1996. "Phonological Change II: Change in Phonological Systems." In: *Historical Linguistics.* London: Arnold. 76–101.

Trudgill, Peter. 1996. *Sociolinguistics: An Introduction to Language and Society.* Revised edition. New York: Penguin.

Uhlirovà, Ludmila. 1994. "E-mail as a New Subvariety of Medium and its Effects upon the Message." In: Svetla Cmejrkovà and Frantisek Stícha (eds.). *The Syntax of Sentence and Text.* Amsterdam, Philadelphia: John Benjamins Publishing Company. 273–282.

Uncle Jazzbeau's Gallimaufrey. Linguistics, philosophy, and politics. <http://www.bisso.com/ujg_archives/000060.html> [12/10/04].

UPOC. North America's premier mobile community. 2005. <http://www.upoc.com/> [15/06/06].

Urban Dictionary. 1999–2006. <http://www.urbandictionary.com/define.php?term =ayt> [05/03/06].

Veredarius, O. [i.e. Hennicke, Ferdinand]. 1885. *Das Buch der Weltpost. Entwicklung und Wirken der Post und Telegraphie im Weltverkehr.* Berlin. Reprint Heidelberg 1984.

Vincent, Jane. 2004. "The Social Shaping of the Mobile Communications Repertoire." In: *The Journal of the Communications*, 1. London: Communications Network. 4–8.

Violi, Patrizia. 1985. "Letters." In: Teun A. Van Dijk (ed.). *Discourse and Literature.* Amsterdam, Philadelphia: John Benjamins Publishing Company. 149–167.

Watts, Richard J. 2004. "Einflüsse des Englischen auf die deutsche Schriftsprache in der Schweiz und die Schweizerdeutschen Dialekte." In: Sandro M. Moraldo and Marcello Soffritti (eds.). *Deutsch aktuell.*

Einführung in die Tendenzen der deutschen Gegenwartssprache. Rom: Carocci. 113–123.

Weingarten, Rüdiger (ed.). 1997. *Sprachwandel durch Computer.* Opladen: Westdeutscher Verlag.

Weingarten, Rüdiger. 2002. "Der Computer als Schriftmuseum. Latinisierung von Schriften durch computertechnische Zwänge?" In: Erika Gerber, Konrad Ehlich and Jan-Dirk Müller (eds.). *Materialität und Medialität von Schrift.* Bielefeld: Aisthesis Verlag. 165–182.

Weininger, Markus J. and Lesley Shield. 2004. "Proximity and distance: a theoretical model for the description and analsyis of online discourse." <http://www.didascalia.be/downloads/CALL2004/PROC-CALL2004_Weininger_Shield.doc> [19/06/06].

Wilss, Wolfram. 2003. "Das Eindringen angloamerikanischer Fremdwörter in die deutsche Sprache seit Ende des Zweiten Weltkrieges (bis 1958)." In: Hermann Zabel (ed.). *Denglish, nein danke! Zur inflationären Verwendung von Anglizismen und Amerikanismen in der deutschen Gegenwartssprache.* 2nd edition. Paderborn: IFB Verlag. 15–32.

Yates, Simeon J. 1996a. "Computer-mediated English: sociolinguistic aspects of computer-mediated communication." In: Janet Maybin and Neil Mercer. *Using English: From Conversation to Canon.* London: Routledge. 76–83.

Yates, Simeon J. 1996b. "Oral and written linguistic aspects of computer conferencing." In: Susan Herring (ed.). *Computer-Mediated Communication: Linguistic, Social and Cross-Cultural Perspectives.* Amsterdam, Philadelphia: John Benjamins Publishing Company. 29–46.

Yongyan, Li. 2000. "Surfing e-mails." In: *English Today,* 64. 16 (4). 31–55.

Zabel, Hermann. 2003. *Denglisch, nein danke! Zur inflationären Verwendung von Anglizismen und Amerikanismen in der deutschen Gegenwartssprache.* 2nd edition. Paderborn: IFB Verlag.

List of Figures

List of Tables

Linguistic Insights

Studies in Language and Communication

This series aims to promote specialist language studies in the fields of linguistic theory and applied linguistics, by publishing volumes that focus on specific aspects of language use in one or several languages and provide valuable insights into language and communication research. A cross-disciplinary approach is favoured and most European languages are accepted.

The series includes two types of books:

- **Monographs** – featuring in-depth studies on special aspects of language theory, language analysis or language teaching.
- **Collected papers** – assembling papers from workshops, conferences or symposia.

Each volume of the series is subjected to a double peer-reviewing process.

Vol. 1 Maurizio Gotti & Marina Dossena (eds)
Modality in Specialized Texts. Selected Papers of the 1st CERLIS Conference.
421 pages. 2001. ISBN 3-906767-10-8. US-ISBN 0-8204-5340-4

Vol. 2 Giuseppina Cortese & Philip Riley (eds)
Domain-specific English. Textual Practices across Communities and Classrooms.
420 pages. 2002. ISBN 3-906768-98-8. US-ISBN 0-8204-5884-8

Vol. 3 Maurizio Gotti, Dorothee Heller & Marina Dossena (eds)
Conflict and Negotiation in Specialized Texts. Selected Papers of the 2nd CERLIS Conference.
470 pages. 2002. ISBN 3-906769-12-7. US-ISBN 0-8204-5887-2

Editorial address:

Prof. Maurizio Gotti Università di Bergamo, Facoltà di Lingue e Letterature Straniere,
Via Salvecchio 19, 24129 Bergamo, Italy
Fax: 0039 035 2052789, E-Mail: m.gotti@unibg.it

Vol. 4 Maurizio Gotti, Marina Dossena, Richard Dury, Roberta Facchinetti & Maria Lima
Variation in Central Modals. A Repertoire of Forms and Types of Usage in Middle English and Early Modern English.
364 pages. 2002. ISBN 3-906769-84-4. US-ISBN 0-8204-5898-8

Vol. 5 Stefania Nuccorini (ed.)
Phrases and Phraseology. Data and Descriptions.
187 pages. 2002. ISBN 3-906770-08-7. US-ISBN 0-8204-5933-X

Vol. 6 Vijay Bhatia, Christopher N. Candlin & Maurizio Gotti (eds)
Legal Discourse in Multilingual and Multicultural Contexts. Arbitration Texts in Europe.
385 pages. 2003. ISBN 3-906770-85-0. US-ISBN 0-8204-6254-3

Vol. 7 Marina Dossena & Charles Jones (eds)
Insights into Late Modern English. 2nd edition.
378 pages. 2003, 2007.
ISBN 978-3-03911-257-9. US-ISBN 978-0-8204-8927-8

Vol. 8 Maurizio Gotti
Specialized Discourse. Linguistic Features and Changing Conventions.
351 pages. 2003, 2005.
ISBN 3-03910-606-6. US-ISBN 0-8204-7000-7

Vol. 9 Alan Partington, John Morley & Louann Haarman (eds)
Corpora and Discourse.
420 pages. 2004. ISBN 3-03910-026-2. US-ISBN 0-8204-6262-4

Vol. 10 Martina Möllering
The Acquisition of German Modal Particles. A Corpus-Based Approach.
290 pages. 2004. ISBN 3-03910-043-2. US-ISBN 0-8204-6273-X

Vol. 11 David Hart (ed.)
English Modality in Context. Diachronic Perspectives.
261 pages. 2003. ISBN 3-03910-046-7. US-ISBN 0-8204-6852-5

Vol. 12 Wendy Swanson
Modes of Co-reference as an Indicator of Genre.
430 pages. 2003. ISBN 3-03910-052-1. US-ISBN 0-8204-6855-X

Vol. 13 Gina Poncini
Discursive Strategies in Multicultural Business Meetings.
2nd edition. 338 pages. 2004, 2007.
ISBN 978-3-03911-296-8. US-ISBN 978-0-8204-8937-7

Vol. 14 Christopher N. Candlin & Maurizio Gotti (eds)
Intercultural Aspects of Specialized Communication.
2nd edition. 369 pages. 2004, 2007.
ISBN 978-3-03911-258-6. US-ISBN 978-0-8204-8926-1

Vol. 15 Gabriella Del Lungo Camiciotti & Elena Tognini Bonelli (eds)
Academic Discourse. New Insights into Evaluation.
234 pages. 2004. ISBN 3-03910-353-9. US-ISBN 0-8204-7016-3

Vol. 16 Marina Dossena & Roger Lass (eds)
Methods and Data in English Historical Dialectology.
405 pages. 2004. ISBN 3-03910-362-8. US-ISBN 0-8204-7018-X

Vol. 17 Judy Noguchi
The Science Review Article. An Opportune Genre in the Construction of Science.
274 pages. 2006. ISBN 3-03910-426-8. US-ISBN 0-8204-7034-1

Vol. 18 Giuseppina Cortese & Anna Duszak (eds)
Identity, Community, Discourse. English in Intercultural Settings.
495 pages. 2005. ISBN 3-03910-632-5. US-ISBN 0-8204-7163-1

Vol. 19 Anna Trosborg & Poul Erik Flyvholm Jørgensen (eds)
Business Discourse. Texts and Contexts.
250 pages. 2005. ISBN 3-03910-606-6. US-ISBN 0-8204-7000-7

Vol. 20 Christopher Williams
Tradition and Change in Legal English. Verbal Constructions in Prescriptive Texts.
2nd revised edition. 216 pages. 2005, 2007. ISBN 978-3-03911-444-3.

Vol. 21 Katarzyna Dziubalska-Kołaczyk & Joanna Przedlacka (eds)
English Pronunciation Models: A Changing Scene.
476 pages. 2005. ISBN 3-03910-662-7. US-ISBN 0-8204-7173-9

Vol. 22 Christián Abello-Contesse, Rubén Chacón-Beltrán, M. Dolores López-Jiménez & M. Mar Torreblanca-López (eds)
Age in L2 Acquisition and Teaching.
214 pages. 2006. ISBN 3-03910-668-6. US-ISBN 0-8204-7174-7

Vol. 23 Vijay K. Bhatia, Maurizio Gotti, Jan Engberg & Dorothee Heller (eds)
Vagueness in Normative Texts.
474 pages. 2005. ISBN 3-03910-653-8. US-ISBN 0-8204-7169-0

Vol. 24 Paul Gillaerts & Maurizio Gotti (eds)
Genre Variation in Business Letters.
407 pages. 2005. ISBN 3-03910-674-0. US-ISBN 0-8204-7552-1

Vol. 25 Ana María Hornero, María José Luzón & Silvia Murillo (eds)
Corpus Linguistics. Applications for the Study of English.
526 pages. 2006. ISBN 3-03910-675-9 / US-ISBN 0-8204-7554-8

Vol. 26 J. Lachlan Mackenzie &
María de los Ángeles Gómez-González (eds)
Studies in Functional Discourse Grammar.
259 pages. 2005. ISBN 3-03910-696-1 / US-ISBN 0-8204-7558-0

Vol. 27 Debbie G.E. Ho
Classroom Talk. Exploring the Sociocultural Structure of Formal ESL Learning. 2nd edition.
254 pages. 2006, 2007. ISBN 978-3-03911-434-4

Vol. 28 Javier Pérez-Guerra, Dolores González-Álvarez,
Jorge L. Bueno-Alonso & Esperanza Rama-Martínez (eds)
'Of Varying Language and Opposing Creed':
New Insights into Late Modern English.
455 pages. 2007. ISBN 978-3-03910-788-9

Vol. 29 Francesca Bargiela-Chiappini & Maurizio Gotti (eds)
Asian Business Discourse(s).
350 pages. 2005. ISBN 3-03910-804-2 / US-ISBN 0-8204-7574-2

Vol. 30 Nicholas Brownlees (ed.)
News Discourse in Early Modern Britain.
Selected Papers of CHINED 2004.
300 pages. 2006. ISBN 3-03910-805-0 / US-ISBN 0-8204-8025-8

Vol. 31 Roberta Facchinetti & Matti Rissanen (eds)
Corpus-based Studies of Diachronic English.
300 pages. 2006. ISBN 3-03910-851-4 / US-ISBN 0-8204-8040-1

Vol. 32 Marina Dossena & Susan M. Fitzmaurice (eds)
Business and Official Correspondence:
Historical Investigations.
209 pages. 2006. ISBN 3-03910-880-8 / US-ISBN 0-8204-8352-4

Vol. 33 Giuliana Garzone & Srikant Sarangi (eds)
Discourse, Ideology and Specialized Communication.
494 pages. 2007. ISBN 978-3-03910-888-6

Vol. 34 Giuliana Garzone & Cornelia Ilie (eds)
The Use of English in Institutional and Business Settings.
An Intercultural Perspective.
372 pages. 2007. ISBN 978-3-03910-889-3

Vol. 35 Vijay K. Bhatia & Maurizio Gotti (eds)
Explorations in Specialized Genres.
316 pages. 2006. ISBN 3-03910-995-2 / US-ISBN 0-8204-8372-9

Vol. 36 Heribert Picht (ed.)
Modern Approaches to Terminological Theories and Applications.
432 pages. 2006. ISBN 3-03911-156-6 / US-ISBN 0-8204-8380-X

Vol. 37 Anne Wagner & Sophie Cacciaguidi-Fahy (eds)
Legal Language and the Search for Clarity / Le langage juridique et la quête de clarté. Practice and Tools / Pratiques et instruments.
487 pages. 2006. ISBN 3-03911-169-8 / US-ISBN 0-8388-5

Vol. 38 Juan Carlos Palmer-Silveira, Miguel F. Ruiz-Garrido & Inmaculada Fortanet-Gómez (eds)
Intercultural and International Business Communication: Theory, Research and Teaching.
343 pages. 2006. ISBN 3-03910-954-5 / US-ISBN 0-8204-8356-7

Vol. 39 Christiane Dalton-Puffer, Dieter Kastovsky, Nikolaus Ritt & Herbert Schendl (eds)
Syntax, Style and Grammatical Norms: English from 1500–2000.
250 pages. 2006. ISBN 3-03911-181-7 / US-ISBN 0-8204-8394-X

Vol. 40 Marina Dossena & Irma Taavitsainen (eds)
Diachronic Perspectives on Domain-Specific English.
280 pages. 2006. ISBN 3-03910-176-0 / US-ISBN 0-8204-8391-5

Vol. 41 John Flowerdew & Maurizio Gotti (eds)
Studies in Specialized Discourse.
293 pages. 2006. ISBN 3-03911-178-7 / US-ISBN 0-8204-8393-1

Vol. 42 Ken Hyland & Marina Bondi (eds)
Academic Discourse Across Disciplines.
320 pages. 2006. ISBN 3-03911-183-3 / US-ISBN 0-8204-8396-6

Vol. 43 Paul Gillaerts & Philip Shaw (eds)
The Map and the Landscape: Norms and Practices in Genre.
256 pages. 2006. ISBN 3-03911-182-5 / US-ISBN 0-8204-8395-4

Vol. 44 Maurizio Gotti & Davide Giannoni (eds)
New Trends in Specialized Discourse Analysis.
301 pages. 2006. ISBN 3-03911-184-1 / US-ISBN 0-8204-8381-8

Vol. 45 Maurizio Gotti & Françoise Salager-Meyer (eds)
Advances in Medical Discourse Analysis:
Oral and Written Contexts.
492 pages. 2006. ISBN 3-03911-185-X / US-ISBN 0-8204-8382-6

Vol. 46 Maurizio Gotti & Susan Šarcević (eds)
Insights into Specialized Translation.
396 pages. 2006. ISBN 3-03911-186-8 / US-ISBN 0-8204-8383-4

Vol. 47 Khurshid Ahmad & Margaret Rogers (eds)
Evidence-based LSP: Translation, Text and Terminology.
584 pages. 2007. ISBN 978-3-03911-187-9

Vol. 48 Hao Sun & Dániel Z. Kádár (eds)
It's the Dragon's Turn: Chinese Institutional Discourses.
262 pages. 2008. ISBN 978-3-03911-175-6

Vol. 49 Cristina Suárez-Gómez
Relativization in Early English (950-1250):
the Position of Relative Clauses.
149 pages. 2006. ISBN 3-03911-203-1 / US-ISBN 0-8204-8904-2

Vol. 50 Maria Vittoria Calvi & Luisa Chierichetti (eds)
Nuevas tendencias en el discurso de especialidad.
319 pages. 2006. ISBN 978-3-03911-261-6

Vol. 51 Mari Carmen Campoy & María José Luzón (eds)
Spoken Corpora in Applied Linguistics.
274 pages. 2008. ISBN 978-3-03911-275-3

Vol. 52 Konrad Ehlich & Dorothee Heller (Hrsg.)
Die Wissenschaft und ihre Sprachen.
323 pages. 2006. ISBN 978-3-03911-272-2

Vol. 53 Jingyu Zhang
The Semantic Salience Hierarchy Model:
The L2 Acquisition of Psych Predicates
273 pages. 2007. ISBN 978-3-03911-300-2

Vol. 54 Norman Fairclough, Giuseppina Cortese & Patrizia Ardizzone (eds)
Discourse and Contemporary Social Change.
555 pages. 2007. ISBN 978-3-03911-276-0

Vol. 55 Jan Engberg, Marianne Grove Ditlevsen, Peter Kastberg &
Martin Stegu (eds)
New Directions in LSP Teaching.
331 pages. 2007. ISBN 978-3-03911-433-7

Vol. 56 Dorothee Heller & Konrad Ehlich (Hrsg.)
Studien zur Rechtskommunikation.
322 pages. 2007. ISBN 978-3-03911-436-8

Vol. 57 Forthcoming.

Vol. 58 Carmen Frehner
Email – SMS – MMS
294 pages. 2008. ISBN 978-3-03911-451-1

Vol. 59 Isabel Balteiro
The Directionality of Conversion in English: A Dia-Synchronic Study.
276 pages. 2007. ISBN 978-3-03911-241-8

Vol. 60 Maria Milagros Del Saz Rubio
English Discourse Markers of Reformulation.
237 pages. 2007. ISBN 978-3-03911-196-1

Vol. 61 Forthcoming.

Vol. 62 Sandrine Onillon
Pratiques et représentations de l'écrit.
458 pages. 2008. ISBN 978-3-03911-464-1.

Vol. 63 Hugo Bowles & Paul Seedhouse (eds)
Conversation Analysis and Language for Specific Purposes.
334 pages. 2007. ISBN 978-3-03911-469-6

Vol. 64–71 Forthcoming.

Vol. 72 Cynthia J. Kellett Bidoli & Elana Ochse (eds)
English in International Deaf Communication.
444 pages. 2008. ISBN 978-3-03911-610-2

Vol. 73 Joan C. Beal, Carmela Nocera & Massimo Sturiale (eds)
Perspectives on Prescriptivism.
269 pages. 2008. ISBN 978-3-03911-632-4

Vol. 74–76 Forthcoming.

Maurizio Gotti / Davide S. Giannoni (eds)

New Trends in Specialized Discourse Analysis

Bern, Berlin, Bruxelles, Frankfurt am Main, New York, Oxford, Wien, 2006. 301 pp.
Linguistic Insights. Studies in Language and Communication. Vol. 44
Edited by Maurizio Gotti
ISBN 978-3-03911-184-8 / US-ISBN 978-0-8204-8381-8 pb.
sFr. 72.– / € 49.60 / €** 51.– / € 46.40 / £ 32.50 / US-$ 55.95*

* includes VAT – valid for Germany ** includes VAT – valid for Austria

This volume brings together a selection of contributions presented at the 15th European Symposium on Languages for Special Purposes, held at the University of Bergamo (Italy) from 29 August to 2 September 2005. The conference title, «New Trends in Specialized Discourse», reflects the emphasis given to recent orientations in research, coming from established as well as new authors in the field.

As suggested by the title of this volume, the analysis of specialized discourse calls for a specialized discourse analysis. When applied linguists deal with vocational discourses, they are faced with a double challenge: on the one hand, an understanding of textualisations often alien to the general language; on the other hand, the use of analytical tools designed specifically for their investigation. The studies presented in this volume position themselves somewhere along this continuum, focusing alternatively on converging/diverging features of texts and discourses.

Contents: Maurizio Gotti/Davide S. Giannoni: Introduction – Christopher N. Candlin: Accounting for Interdiscursivity: Challenges to Professional Expertise – Marianne Grove Ditlevsen: Der Geschäftsbericht als komplexe Textsorte – Inger Askehave: Spirituality in the Corporate World – Patrick Leroyer/Bernt Møller : Intégration du genre dans le dictionnaire d'entreprise : Le cas des *case stories* – Donatella Malavasi: Evaluation in Banks' Annual Reports: A Comparison of EL1 and EIL texts – Diana Yankova: The Interplay of Structural and Semantic Constituents in Statutory Texts – Stanisław Goźdź-Roszkowski: Frequent Phraseology in Contractual Instruments: A Corpus-based Study – Cecile Desoutter: Le processus de simplification du langage administratif en France – Aino Piehl: The Influence of EU Legislation on Finnish Legal Discourse – Khurshid Ahmad: Metaphors in the Languages of Science? – Susanne Göpferich: How Comprehensible are Popular Science Texts? The Use of Thinking-aloud Protocols and Log Files of Reverbalization Processes in Comprehensibility Research – Estrella Montolío Durán: Formación en técnicas de comunicación escrita dirigida a ingenieros de informática. El caso de la marcación de la operación textual de la enumeración – Paola Catenaccio: The SARS Crisis Coverage in the *British Medical Journal* and *New Scientist.*

PETER LANG
Bern · Berlin · Bruxelles · Frankfurt am Main · New York · Oxford · Wien